No 39

NATIONAL FORESTS AND RECREATIONAL OPPORTUNITIES

BRIAN GOODALL

geographical papers

Reading Geographical Papers
Department of Geography
University of Reading
Whiteknights Reading England

Published November 1975

© Brian Goodall

ISBN No 0 7049 0346 6

ISSN No 0305 5914

Editors

Michael Batty   Ronald Botham   Mark Ebery   Michael Wooding

Assistant Editor

Brian Preston

Printed by George Over Ltd., London and Rugby

CONTENTS

	Page
1. COUNTRYSIDE RECREATION AND THE ROLE OF FOREST	1
Increasing Recreational Demands	1
The Changing Countryside	1
Focus on Forests	2
2. FORESTRY COMMISSION POLICY AND CURRENT RECREATIONAL USAGE	5
Development of Recreational Policy	5
The Present Policy	6
The Facilities Currently Available	8
Current Levels of Recreational Use	11
Day visitors	11
Overnight visitors	12
Specialist uses	12
Educational uses	14
The Value of Forest Recreational Use	15
3. FOREST RECREATIONAL POTENTIAL	16
The Suitability of Forest for Recreation	16
The Supply of Forest Resources for Recreation	20
The size of the Forestry Commission estate	20
The regional distribution of national forests	21
A qualitative view of national forests	24
Measuring the Recreational Quality of a Forest Environment	28
The Demand for Forest Recreation	32
4. PLANNING OF FOREST RECREATION: PROBLEMS AND PRINCIPLES	35
Planning for Recreation: Forest Versus Non-Forest Environments	35
Multiple Use - A Guiding Principle in the Recreational Use of Forest?	37
The Compatibility of Recreational Activities in a Forest Environment	41
Some Principles for the Recreational Use of Forests	44
5. A RECREATIONAL STRATEGY FOR NATIONAL FORESTS	48
Resource Availability and Recreational Strategy	48
A Strategic Spatial Framework for National Forest Recreation	48
The Spatial Implications of the Overall Strategy in Depth	51
Highland forests	51
Lowland forests	52
Urban forests	53
Conclusion	53
REFERENCES	55

ABSTRACT

   Forest could become an increasingly important recreational resource base in a changing countryside and the existence of national forests may prove to be a vital asset. The development of recreational policy towards national forests and the current levels of usage and facilities available are surveyed. Discussion then centres on the suitability of forest for recreation and the supply of national forest resources. Multiple use and recreational compatibilities are reviewed as guiding principles of recreational planning before suggesting a spatial, strategic framework for national forest recreation.

ACKNOWLEDGEMENT

   I would like to thank Dr. J.B. Whittow for his comments on an earlier draft of this paper and for the stimulus which derived from working with him on a research project concerning the recreational potential of Forestry Commission holdings. I would also like to thank H. Walkland, Department of Geography, for photoreduction of the maps, and Mrs. J. Preston for the typing.

   Brian Goodall is a staff member in the Department of Geography, University of Reading.

1. COUNTRYSIDE RECREATION AND THE ROLE OF FOREST

Increasing Recreational Demands

The continuing growth of outdoor recreation must be viewed as an expression of a steady change in British leisure pursuits. In this context the relative importance of a rural environment for the recreational activity of a predominantly urban population is striking and suggests the characteristics of that environment have a positive contribution to make to the recreational experience. It should not be inferred from this, however, that recreation is a new countryside phenomenon but simply that the incursion of urban dwellers is on a massive scale. There is general agreement about the main factors leading to this increase: the widening acceptance of a standard five-day working week and of appreciably longer holidays with pay give people more time to visit the countryside; rising real incomes leave people with more money to spend after basic necessities have been met; the immense growth in private car ownership has revolutionized the general mobility of the population; and better education has stimulated the positive use of leisure time.

It is generally expected that recreational demands on the countryside will continue to increase, although the rate and extent of that increase can only be loosely inferred from currently available data. Not only will a greater quantity of countryside be sought for recreation but there will also be an increasing concern for the quality of that recreational environment. Government policy, written into the 1968 Countryside Act, is to encourage provision for recreational needs in the countryside. But where can, and should, that provision be made in the countryside?

The Changing Countryside

The countryside is not unchanging and increasing recreational use is not the only change to be accommodated. The greatest changes are within agriculture, especially in the lowlands, where a combination of mechanized systems of cropping necessitating larger fields and intensive livestock rearing processes requiring large buildings and controlled grazing has led to the gradual disappearance in many areas of the traditional, small scale, intimate English landscape. Agricultural change in hill farming areas has been less dramatic but even here farm units are undergoing structural changes and poorer land is being abandoned. These changing agricultural practices render the countryside less desirable for much recreational activity and increase the likelihood of conflict between agricultural and recreational interests. Recreation is therefore being increasingly confined, particularly in lowland areas, to non-agricultural land. In those lowland areas where commons are few and wild areas non-existent woodland may represent the only space available for outdoor recreation.

The rising demand for recreation may be met by a combined course of action which increases the capacity of existing recreational resources, creates new resources for recreation and develops multiple use of rural land. Few rural resources are devoted entirely to recreation. Coppock (1968) estimated that land used primarily for recreation - other than general open space - accounted for less than 1% of all rural land. Moreover, many of the traditional countryside recreation areas, such as the New Forest, may well be suffering from saturation and therefore offer limited prospects for increasing capacity. Opportunities for creating new recreational resources are infrequent although the conversion of worked-out wet gravel pits to country parks indicates it is not unknown. As noted above changing agricultural practices are likely to reduce the possibility of the joint use of land for recreation. Forest or woodland may offer the best opportunities for expanding rural recreational capacity both in terms of providing new resources and of improved integration of recreation with commercial forestry. Forests probably form the largest part of rural non-agricultural land and, if the general public turns to forests as in the United States to satisfy much of their recreational needs, they are likely to bear the major brunt of the rising recreational demands.

Focus on Forests

Since outdoor recreational activities are not unique to forests and can usually be carried out equally well, though with different emphasis, on other areas of countryside, forest recreation is just one aspect of countryside recreation. Forests, however, have a positive contribution to make to recreation because not only do they offer facilities for the quiet enjoyment of air, exercise and scenery but they also provide sites for the practice of the more noisy or visually unattractive recreational pursuits. The recreational use of forests may therefore be increasingly desirable for social reasons. In addition the traditional field sports and shooting interests still value forest highly. Forest and woodland therefore appear as a major recreational asset.

In comparison with most other advanced countries Great Britain is lightly wooded with only 8% of its total land area wooded (cf. France 21%, West Germany 29%, U.S.A. 33%, Canada 48%, Sweden 57%) and this represents only 0.03 hectares per person compared to a world average of 1.18 hectares per person (Forestry Commission, 1974). The current pattern of forest distribution is a consequence of centuries of deforestation by grazing, burning and clearance for agriculture, timber and fuel, which has left little natural forest cover, and of the subsequent afforestation policy of the last fifty years which has resulted in the formstion of sizeable commercial plantations.

The possible extent of forest recreational resources is indicated in Table 1. This shows that, in 1973, there were nearly 2 million hectares of forest in Great Britain of which 47% was in England, 41.5% in Scotland and 11.5% in Wales. Some 81.5% of this hectarage is classified as productive forest and this, in terms of ownership is divided almost equally between the state and private owners. The 18.5% of

TABLE 1: FOREST AREA BY OWNERSHIP AND USE, 1973[a]
(Areas in 000 hectares)

	Productive Forests			
Region	Forestry Commission	Private	Unproductive[b]	Total
England	240	476	188	904
Scotland	400	264	135	799
Wales	129	60	33	222
Great Britain	768	800	356	1924

Source: Forestry Commission (1974) British Forestry.
a. Totals do not always add because of rounding
b. Largely (93.25%) privately owned. The small proportion in Forestry Commission ownership often represents land acquired from private ownership for restocking.

unproductive forest, defined as woodland which is unlikely to develop a satisfactory crop, is overwhelmingly in private ownership. The area under forest is now more than one-third greater than in 1913 and one-fifth greater than in 1947. This increased forest hectarage is largely the result of government policy designed oroginally to create a strategic reserve of timber. This reserve was to be achieved as a result of the activities of both a state forestry concern - the Forestry Commission - and private timber growers. Various targets were envisaged for this reserve but it was generally agreed after World War 2 to aim for 2 million hectares of productive forest by the end of the century. Three-fifths of that target area would be achieved by afforestation. The net result of this policy to date is that the nation has acquired, especially in the upland areas of Great Britain, substantial areas of publicly owned forest.

Since private forests are often smaller units and not generally open to the public in the sense that most Forestry Commission ones are and because the use of private land for forestry was often encouraged by the attractive tax concessions available, private woodland is generally less readily available for recreational use and their owners frequently less interested in promoting recreational activity. National or state forests constitute the largest area of rural land within unified public ownership and, managed and planned as such, represent potentially a major recreational asset. Indeed the Forestry Commission has already shown limited recreational activities to be compatible with its other interests and has proved forests to be eminently suitable for these activities. However there is ample scope for the intensification of existing and the introduction of new recreational activities into national forests.

The introduction of recreational activity into a forest may be regarded as an innovation and the public-private dichotomization of forest land ownership noted above may be related to the adaptor-adoptor concept in the diffusion or spatial acceptance of an innovation. The Forestry Commission may be regarded as the initiator or adaptor by being the first to promote recreational activity in a forested environment. Some private forest owners will follow a successful Forestry Commission lead and these will be the adoptors of the innovation. Since a large proportion of the recreational resources likely to be needed to meet increasing recreational demands are under public control and since private initiative on any scale seems less likely it may be argued that the government should encourage the Forestry Commission to take further action to realise the recreational potential of its forests for the benefit of the nation. A spin-off, in the form of increased availability and development of private forests for recreation, must also be hoped for.

2. FORESTRY COMMISSION POLICY AND CURRENT RECREATIONAL USAGE

Development of Recreational Policy

The Forestry Commission was formed in 1919 for the specific task of building up a strategic reserve of standing timber to make good the depletion of the nation's timber resources during the First World War. The Forestry Commission acts as the Forest Authority in promoting the general interests of forestry, developing afforestation, encouraging and guiding private forestry and helping maintain an efficient timber trade. It also functions as the Forest Enterprise in establishing and maintaining national forests and marketing the products from those forests. The concept of a strategic reserve remained at the core of forestry policy until the late 1950s and although timber production remains the primary objective of the Forestry Commission's estate, other uses, including recreation, are encouraged in line with the theory of multiple use. The Forestry Commission undertook, in the 1930s pioneer development in the field of outdoor recreation when, inspired by the call for National Parks, it established a number of Forest Parks. The post-war policy statement (Forestry Commission, 1943) recommended more attention be given to public access and the provision of recreational opportunities in forests, especially by the formation of more Forest Parks. By the early 1960s the success of Forest Parks and the realisation of the impressive recreational benefits derivable from forests encouraged the Forestry Commission to open up other forests by providing improved access and simple facilities as a response to the increasing number of visitors to the countryside. Financial considerations restricted the scope of positive action but the concept of 'Open Forests' evolved. During this period there was much well-intentioned compromise resulting from the efforts, often the work of enthusiastic individuals, to accommodate recreational demand as it appeared at the local level, that is, the individual forest. Little attempt, however, was made to establish a comprehensive national approach to recreation which was fully integrated with the Commission's other objectives.

Formal powers to develop the recreational potential of national forests by providing tourist, recreational and sporting facilities and to make charges for those facilities were not vested in the Forestry Commission until the 1967 Countryside (Scotland) Act and the 1968 Countryside Act. The role of forests in national recreational strategy is thus recognised although the Countryside Commission cannot directly aid development on Forestry Commission land. The Countryside Commission may, however, grant aid to local authorities who wish to acquire and manage sites on lease from the Forestry Commission. In addition these Acts extend the powers of the Forestry Commission to include the planting, care and maintenance of trees in the interests of amenity and the purchase of land for such purposes. The initiative therefore rests with the Forestry Commission. In 1970 the Commission established, as part of its headquarter's organisation, a Conservation and Recreation Branch whose first task was to formulate a policy for the development of the recreational potential of national forests in the light of the Commission's wider recreational interests and responsibilities. To aid the implementation of the policy each conservancy appointed a recreation planning officer to co-ordinate developments at the regional level. Recreational plans for each conservancy are now in hand. The 1972 forestry policy review (Ministry of Agriculture, 1972) also declares itself to be in favour of increased recreational management. The framework for a comprehensive approach to national forest recreation has therefore been established but, although visitors are welcomed and public demand for facilities is catered for where it makes itself felt, little initiative has been taken to generate additional demand or consider the widest possible range of recreational activity.

The Present Policy

The progressive policy of increasing recreational provision in national forests has been influenced not only by the increasing demand for countryside recreation but also by the realisation of the small contribution currently made by forests to the satisfaction of that demand and by recognition of the low costs incurred in adapting forest management to accommodate recreation and of the high benefits derivable from informal recreation in a forest environment. The belief that the majority of people visiting the countryside simply want a place to park, to picnic and to walk underlies the policy adopted. The course of action which follows fits conveniently with multiple use theory where timber production is the dominant use and public access for recreation is allowed, along with the provision of appropriate facilities, so as not to prejudice the forest environment or involve major capital expenditure or detract significantly from the main product. Justification for such a policy also rests on the idea that the recreational potential of national forests is being realised for the benefit of the majority of society and priorities are established accordingly.

The Forestry Commission's general recreation policy, as formally stated in its fifty-first annual report (Forestry Commission, 1971A), is "to develop the unique recreational features and potential of its forests, particularly where they are readily accessible to large numbers of visitors from the major cities and holiday centres. This will be done in conformity with the Commission's statutory powers and obligations, within the financial resources available and subject to the primary objective of timber production. The Commission will ensure that its recreational development will neither injure the forest environment nor conflict with its conservation."

Where any conflict between recreational activities is likely to occur the aim is to give priority to the informal recreational needs of the general public. However that priority is constrained to the extent that the Commission's policy is to allow the public access only on foot to any of its forests, except those where unrestricted access conflicts with management requirements or obligations to lessors and tenants. Forests are therefore seen to have a key role to play as areas where people can escape from cars. Any provision for the motorist, such as parking facilities, will be confined to the forest edge. The policy, in addition, seeks to maintain the status of the Forest Parks and to encourage the widest use of forests for educational purposes and the study of natural history.

As a general rule the public is allowed to use national forests for scenic enjoyment, air and exercise, along with simple basic facilities such as picnic places and forest walks, free of charge. This stems, as much as anything from established custom since the many small areas used by forest visitors adjoin public roads, are scattered and have been freely used for many years prior to their formal development. Certain other facilities - some car parks and information centres - will not be expected to be self-supporting and an 'honesty box' system may be used to collect nominal charges. Where larger facilities are provided, like sites for camping and caravanning, large scale car parks and information centres, charges will be made since these activities are required to be self-supporting and a net return of 10% on capital invested is generally expected.

The public's use of national forests is subject to the standard bye-laws of the Forestry Commission which are designed to protect the trees and their environment and safeguard that use and enjoyment against the careless or malicious behaviour of a very small minority of the public. These bye-laws came into effect from 1st July 1971 (Forestry Commission, 1972). Where consideration is given to the possibility of development being carried out in Forestry Commission land by, and either wholly or partly financed by, local authorities and recognised recreational organisations the Forestry Commission reserves the necessary control to co-ordinate that development with its other management objectives.

In respect of specialist activities the Commission's policy is to continue to manage facilities for field sports in accordance with accepted codes of practice subject to general recreational needs being given priority where conflict arises and legal circumstances permit. Opportunities for fishing will be extended and managed to increase the availability of fishing on a daily basis. For other specialist activities, such as car rallying and horse riding, use will depend on the issue of permits and arrangements are made in consultation with representatives of the particular activities. In these cases the rule is to obtain sufficient revenue from the activities concerned to cover any costs incurred by the Forestry Commission.

Thus the Commission's current recreation policy sees the role of forests in the national context as providing opportunities to meet the less specialised, informal recreational demands of the general public. The priority to be given to developing the recreational potential of a given forest is seen as being dependent on the proximity of that forest to large centres of population, including holiday areas. Emphasis is therefore being laid on the quiet enjoyment of fresh air and exercise: there is no intention to provide facilities for urban amusements or stadia for team games. Forestry Commission management aims at flexibility and extremes are avoided in order to keep a range of options open thus the recreational activities favoured are the least disruptive to production forestry and conflict within that range of recreational activity can be resolved by time and space zoning. Recreation plans currently being formulated by the Commission will be designed to make the provision of facilities more deliberate and systematic and, within a budget constraint, investment will be encouraged in facilities for the range of activity discussed above which maximises the excess of incremental benefit over incremental cost (Grayson, et al, 1973).

The Facilities Currently Available

The only forests which have been singled out as having an explicit recreational function are those in areas designated as Forest Parks. Their location is shown on Fig. 1 and details of their area and year of designation are given in Table 2. The New Forest, comprising a further 25,000 hectares of woodland and heathland, although not designated as a Forest Park has been managed as such since the Forestry Commission took over management of the Crown Woods in 1924. The Forest Parks have normally been established in areas where open, mountainous or wilderness country has been acquired in connection with afforestation. All the parks have at least one first-class site for camping and caravanning, car parks and picnic places adjoining the public road system, a network of routes for walkers and riders and most have, or will soon have, a forest interpretation and information centre. Besides the attraction of the open areas for the hill walker these parks may also offer particular opportunities, such as mountaineering, canoeing and sailing in Argyll and Glenmore parks and skiing in Glenmore.

TABLE 2: FOREST PARKS

Park	Area (hectares)	Year Designated
Galloway	52,600	1943
Border	51,000	1955
Argyll	25,500	1935
Queen Elizabeth	20,200	1953
Dean	14,200	1938
Snowdonia	8,000	1937
Glenmore	2,600	1948
	174,100	

Source: Forestry Commission (1974) and Edlin (1969A)

The facilities specifically provided by the Forestry Commission, such as car parks and forest trails, are primarily for informal recreation use. They are generally dispersed, except for overnight accommodation, throughout the Commission's estate (see Fig. 1) and now about two-thirds of the national forests have, at least, a car park. This dispersion spreads

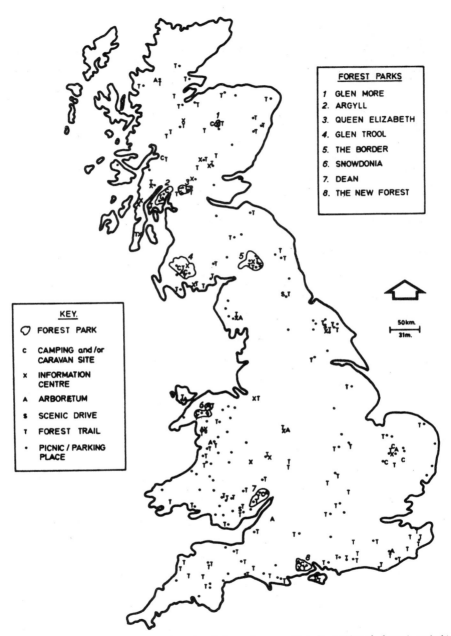

Fig.1 Distribution of facilities for informal recreation in national forest estate
Source: Forestry Commission (1973)

visitors and helps ensure they find conditions for the quiet enjoyment of the forest environment as well as minimizing the likelihood of conflict with timber production in any given location. The pattern of occurrence also reflects the casual development of facilities dependent, in the past, upon local initiative or the existence of well known beauty spots.

In the five years ending 1974 the Forestry Commission nearly doubled the number of car parks available and these now total some 200; the number of picnic places provided more than doubled, to over 300; and the number of forest trails more than trebled, to nearly 400 (Forestry Commission, 1974 and 1975). The car parks are usually small, catering for between 15 and 20 vehicles at a time, whilst the picnic places are relatively simple, being equipped with combined table/benches. Forest trails, of varying length but generally between 1½ and 3 km., are way-marked routes, usually starting from a car park or picnic place, for which self-guide, instructional pamphlets are normally available. To date five scenic forest drives on forest, not public, roads have been opened but, apart from Cwmcarn, they are all in North-east England. The Cwmcarn drive, opened in May 1972, is an 11 km. route and was used by some 17,500 cars (at 30p per car) during its first season (Forestry Commission, 1973). There are now 24 forest information centres, of which nearly a third are in the Forest Parks, with displays and exhibits which help visitors to understand and recognise what is going on around them. Some 26 observation towers are available to the public from which they can watch wildlife. In addition there are the arboreta, or collections of specimen forest trees, which are open to the public, the most notable of which are the National Pinetum at Bedgebury, Westonbirt Arboretum, and Crarae forest garden. An unusual, perhaps unique, development in Britain is the 230-seat open-air theatre in Grizedale Forest which is used for a range of interests from natural history lectures to music recitals. This, along with the co-operative development, by the Forestry Commission and Hampshire County Council, of a country park in the Queen Elizabeth Forest and at adjacent Butser Hill could open the way for other interesting future possibilities in forest recreation.

Provision of facilities for overnight accommodation require the grouping of units within a given forest as well as showing considerable concentration at the national level (see Fig. 1), since, of the 16 camping and caravanning sites now available 75% are within Forest Parks (including the New Forest) and the remainder are in the forests of the Thetford Chase area. These sites, each under the care of a resident warden, have good access roads, hard standings for cars and caravans, and are equipped with flush toilets, hot water and showers. They are normally open from the beginning of April until the end of September. The management of these sites is kept firmly in Forestry Commission hands and is not let to concessionaires. Indeed the Forestry Commission's efforts are a model for other forest landlords since the provision of camping facilities is generally sensitively done, maintaining a high standard of site landscaping. These camp sites, together with forest cabins (as at Strathyre) offer holiday accommodation for about 4,300 family units (Forestry Commission, 1974). In addition the Forestry Commission provides, in co-operation with youth organisations, reserved camp grounds (which are little more than grassy meadows with a piped water supply laid on) and adventure centres, most of which are in or near the Forest Parks and has leased sites for development by organisations such as the Caravan Club, for example, at Barton Mills, Kielder, and Swaffham.

The Forestry Commission is currently investing over £1 million annually in the provision of facilities for informal recreation and overnight accommodation. For the future the Commission plans to increase provision of the types of facilities discussed above, doubling its camp sites within a decade, opening more information centres and continuing to improve access to national forests by providing more car parks, picnic places and forest trails.

Current Levels of Recreational Use

The current recreational usage of national forests may be outlined under four heads: day visitors enjoying informal or passive recreational pursuits; overnight visitors; specialist uses, such as car rallying and shooting; and organised educational purposes.

Day visitors

Not surprisingly informal use by day visitors is the most common recreational use of forests since so few specific facilities are provided for anything else. Although the problem of determining both the number of countryside visitors and of those to forests is immense it has been estimated that Forestry Commission land accounts for about 10% of the total number of day visits to the countryside (Grayson et al., 1973). A sample survey was conducted of day visitors to Forestry Commission land for the period from 1st June to 30th September, 1968 and distinguished the 'concentrated use' sites (defined as sites where over 15 cars were parked on a Sunday afternoon) from the 'lesser use' sites (with between 5 and 15 cars parked) and the 'linear sites' (where at least 4 cars were parked in a 0.4 km stretch along a public road). The results of this survey are summarised in Table 3 along with figures for the New Forest and the Forest of Deen. The total of 15.5 million

TABLE 3: ESTIMATED DAY VISITORS TO FORESTRY COMMISSION LAND, 1968

Area	Seasonal Total of Day Visits (in millions)
Concentrated use sites	6.0
Lesser use sites	0.6
Linear use sites	6.7
New Forest	1.8
Forest of Dean	0.4
	15.5

Source: Forestry Commission, 1970

day-visits excludes visitors to the arboreta and is accurate within ± 5 or 6 million, giving an implied visitor-hours total, at an average of 1.3 hours per visit, of between 14 and 28 million hours (Grayson et al., 1973). A survey for the year ending 30th September, 1969 estimated 3.7 million day visits were made to the New Forest, 79% of them between April and September and 35% between mid-July and the end of August (Forestry Commission, 1969).

About 90% of these day visitors to national forests come by car: amongst these many are families who come on Sunday afternoons during the summer, there is an over representation of the middle income groups, and there is a general reluctance on the part of many visitors to stray far from their cars. The investigation of recreational usage of Cannock, Allerston, Glen More and Trossachs forests sponsored by the Forestry Commission showed picnicking, pleasure driving, walking and climbing to be the activities most commonly undertaken by visitors, although participation at Glen More reflected its greater distance from large cities and the fact that it was a holiday area in its own right (Mutch, 1968). Three-quarters of the visitors in these cases professed to being aware of the opportunities available before making their visit. The New Forest survey (Forestry Commission, 1969) also found picnicking, pleasure driving and strolling, along with informal games, in that order, to be the most popular activities amongst day visitors. In the Forest of Dean a difference was noted between the use of beauty spots by long-trip, all day visitors and dispersed recreation areas by short-stay, local residents (Colenutt and Sideway, 1973).

Overnight visitors

Overnight visitors are almost exclusively caravanners and campers since little other accommodation, in the form of cabins or cottages to let, is currently available. Compared to the total British holiday market it is estimated that holidays in forests account for under 0.5% of all holidays (English Tourist Board, 1972). In terms of the total number of camper nights spent at major Commission camp sites there has been a 218% increase over 8 years to give a 1973 total of 1.2 million camper nights (Forestry Commission, 1974). This increase partly reflects the opening of additional camp sites, three since 1972, and partly the greater number of persons participating in camping and caravanning which allows a higher ratio of filled to available pitches to be maintained throughout the season. The total probably represents little more than 2% of all camper nights in Great Britain at the present time.

Specialist uses

Under one-fifth of visitors to national forests come to participate in a particular activity (Lloyd, 1972). The specialist recreational uses of forest may be discussed in terms of three groups: the traditional use for field sports; specialist uses, such as orienteering, encouraged by the Forestry Commission; and other activities, such as car rallying, which can be accommodated in national forests.

The use of forests for hunting, stalking and shooting is considered as a separate group because sporting rights are a valuable commodity and the terms under which they are available frequently place severe limitations on other recreational use of the forest. There are currently some 2,300 lettings of sporting (including fishing) rights (Edlin, 1969B). In the past where the Forestry Commission obtained land on leasehold, the freeholders normally reserved for themselves the sporting rights. Today this reservation appears as an effective barrier to the Commission's ability to open certain forests to the general public. In other cases, where the sporting rights are controlled by the Commission these have been leased, where appropriate, to a sporting tenant. The leasing of sporting rights by the Commission has always been within a conservation framework designed to link control with protection. Moreover, the Commission will not now renew a sporting lease where it considers it desirable to favour public access. Thus, in the case of deer, which are widespread in British forests and must be controlled particularly in new plantations which are their preferred habitats, stalking is confined to a small part of the Commission's estate, taking place in about 14% of its forests. Even so roe deer stalking has been attracting continental visitors since it is up to 200% cheaper here than on the continent, with the culling of the bucks being particularly attractive to European trophy hunters. Whilst shooting cannot be made available by lease to the general public some consideration is now being given to making it available on a day-permit basis. However, tapping of further opportunities is likely to be limited because of the strict United Kingdom firearms control and the Forestry Commission's understandable insistence on a high standard of marksmanship and requirement that each stalker is accompanied by a Commission ranger. Experiments are currently underway in forests of the Southern Uplands of Scotland to provide pheasant shooting on a day-permit basis.

Of the special activities encouraged by the Forestry Commission natural history pursuits, fishing, orienteering and pony-trekking are prominent. With fishing, the object is to let the general public have access to any opportunities and, where convenient, fishing rights are retained by the Commission and made available on a daily basis. Otherwise the policy is to let waters only to angling societies which allow members of the public to fish. Availability of fishing on Commission land is largely a matter of chance, depending on whether lakes, streams and rivers are acquired when obtaining land for afforestation. Currently only 14% of national forests offer opportunities for fishing, although, with a stocking and improvement policy further waters may be developed. As little as 1% of the total area of coniferous woodland is managed for wildlife per se (Grant, 1971). Even so forests provide, anyway, habitats for a wide variety of wildlife, the existence of which may add to the enjoyment of informal recreational activities, as well as providing a recreational base for wildlife study in its own right. Over half - 53% - of national forests staged orienteering events in 1972-1973 (Forestry Commission, 1974). Orienteering favours forest environments because they offer a greater test of competitors' skills for the restricted visibility forces each cometitor to place greater reliance on the accurate use of compass and map and reduces the likelihood of his being influenced by the routes taken by rivals.

A further group of activities, accommodated by the Forestry Commission sometimes on a relatively large scale, are not actively promoted or sought either because facilities occur by chance on Commission land (as for rock climbing in 15 forests, potholing in 4 and skiing in 1) or because there is pressure on the activities themselves to find socially acceptable locations at which to stage events. The latter pressure derives from potential incompatibility with other activities, not only recreational ones. Car rallying is the most prominent of such activities but horse-riding may also be included. Some 55% of national forests were used to stage car rallies in 1972-73. This is primarily a winter activity and is controlled by national agreement with the Royal Automobile Club. Each forest used, not necessarily the same ones each year, stages a maximum of 2 or 3 rallies per year. Car rallying finds national forests attractive for its purposes on two counts: on the one hand its organisers are under pressure from police forces to get rally events off public roads and on the other hand the popularity of rallying, and especially the international reputation of the R.A.C. Rally of Great Britain, stems from the nature of the special stages which can be promoted on Commission land using the gravel-surfaced forest roads. However in the case of the R.A.C. Rally the cost of using Commission roads - 15p per starting car per 1.6 km of forest road - forces the organisers to seek special stages outside forests but these are regarded as 'Mickey Mouse' stages by the competitors (G.P., 1973). Nearly one-third of forests are used for horse-riding which is controlled by the issue of permits, often to local riding schools, since horse-riding conflicts with many other recreational activities and must be confined to particular trackways in the forest. Such facilities are in addition to the use of public bridleways which transect Commission land.

Educational uses

In addition to encouraging the use of forests for wildlife study by the general public and recognising the educational value of information centres established primarily for the general visitor the Forestry Commission makes specific provision for the reception of organised educational and scientific parties. Where complete ranges of species or particular habitats of one or more rare species of flora and fauna occur on Commission land these areas may be designated, if of national importance, as Forest Nature Reserves or Sites of Special Scientific Interest and managed in agreement with the Nature Conservancy. Areas of lesser importance may be managed directly or jointly with county naturalists' trusts or in agreement with the Nature Conservancy. The Commission also adopts a positive policy of co-operation with primary and secondary schools interested in mounting forestry projects. Some 120 schools now participate in the School Forest Plot scheme.

The Value of Forest Recreational Use

As noted above the Forestry Commission does not seek to maximize the income derived from the recreational activities using its forests and deliberately provides many facilities for informal, day-visitor recreation free of charge. In other cases charges are set at levels which cover basically administrative and supervisory costs and, in yet other cases to show a 10% return on the investment. The total revenue received by the Commission during 1970-71 from its recreational activities was of the order of £0.25 m, of which 40% arose from specialist activities and as much again from camping (Grayson et al., 1973). However considerable difficulty is experienced in attempting to establish an overall value which includes the day-visitor activities. The interdepartmental cost/benefit study (Ministry of Agriculture, 1972), using the Clawson method and taking a value of £0.05 per visitor-hour, estimated the consumer surplus attaching to day-visits in 1968 to be around £1.1 m and giving, when the consumer surpluses for camping and specialist activities were included, a total value of £1.5 m. If the benefit from day-visiting is related only to the hectarage of trees over 25 years of age - some 170,000 hectares - the average surplus from day-visits is £1.0 per hectare for Great Britain, ranging from £0.16 per hectare in East Scotland to £3.5 per hectare in South-East England. (Grayson et al., 1973). Such intra-conservancy variation derives more from the differences in accessibility of Forestry Commission estates to centres of population than to hectarages available.

The social benefits derived from the recreational use of national forests are put into perspective when it is noted that the 1968 revenue from sales of wood by the Commission amounted to only £2.5m. By the mid-1980's the cost/benefit study estimates that the value of the Forestry Commission estate as a recreational asset will equal its value as a source of timber. It should also be noted that the consumer surplus value is based only on actual recreation visits to Commission Forests and does not include any evaluation of benefit in respect of recreationists, such as pleasure motorists, who derive benefit from the scenic or amenity contributions of forest (Duffield and Owen, 1972).

3. FOREST RECREATIONAL POTENTIAL

The Suitability of Forest for Recreation

Current usage already indicates the suitability of forest as an outdoor environment for a range of recreational activities. The benefits enjoyed by participants in outdoor recreational activities depend, in varying degree, on the character and quality of the background landscape. Such considerations are important for nature-based and many informal recreational activities but less so for most active outdoor sports and least of all for spectator sports. In a forest environment the landscape is a function of terrain (physical site) conditions and mantle characteristics and this distinction is of significance for forest recreation (Goodall and Whittow, 1973 and 1975). Recreational benefits may be derived from the practice of an activity in a variety of locations where certain basic site requirements are met. Since these benefits are not unique to forest its recreational use is, therefore, dependent on mantle features of positive recreational interest. It is not however merely a matter of deciding which recreational activities can or cannot take place in a forest environment since many activities are flexible enough to make use of a variety of forested and non-forested environments. Whilst emphasis will here be placed on outdoor recreational activities this does not preclude the possibility of buildings to cater for indoor urban pastimes and stadia for team sports from being erected in a forested area.

Physical factors considered important by recreational activities themselves in influencing the suitability of a site for recreation are climate, space, slope, dissection, ground texture, local accessibility and presence of water (Goodall and Whittow, 1973). These factors apply equally to forested and non-forested sites. Whilst different recreational activities may seek different basic conditions they may still be influenced by a common factor. Climate is particularly important in explaining the marked peaking of informal recreational activity during the summer months. In the case of many activities the availability of areas of open space, that is non-planted land, is essential. Critical slopes or gradients may be established for many activities and generally the number of activities practised decreases with increasing steepness of slope. Local dissection or undulation adds interest to the landscape for informal recreation and is also a requirement for activities such as cyclocross. In respect of ground surface texture, activities normally require a surface passable to either vehicles or to persons on foot or horseback, with a strong preference being expressed for dry firm ground with low vegetation. Local accessibility, interpreted as distance from a motorable road, is important for most recreational activities. The presence of water, depending on its extent and quality, opens up prospects for a range of water-based activities as well as enhancing the recreational environment for campers, caravanners and day-visitors.

Six mantle characteristics, which serve to repel or attract recreational activities, emerge as significant: they are internal forest layout, age and height of trees, penetrability of plantations and spacing of trees, variety of tree species, a screening factor, and the nature of the forest route network. Since few recreational activities make use of totally wooded areas the layout of a commercial forest, particularly the size and distribution of open or non-planted land, is critical. Height of trees proved to be a most important factor because of its influence on the vertical component of landscape. Tree spacing and plantation penetrability are positively related to tree height although the forest management practices to control the development of ground vegetation are significant. Recreational activities appear more concerned with species diversity per se than with the broadleaf-conifer distinction. The screening advantages of forest - the reduction of visual intrusion, absorption of noise, creation of shade and shelter - were appreciated by all recreational activities but particularly by those activities aware of the degree of disturbance or benefit reduction (that is negative externality effects) their practice imposes on other activities. The integrated route network which is developed in a commercial forest proves very attractive to 'linear' activities such as horse-riding and car rallying. It must also be noted that many of the mantle characteristics are dynamic and both the range of possible recreational activities and the degree of intensity with which certain activities are practised increases markedly as trees mature.

A study of the minimum or threshold requirements of recreational activities revealed a need for a wide variety of basic physical or site conditions (Goodall and Whittow, 1973 and 1975). Absence of the required conditions in a forest will preclude the practice of specific activities there. However, unattractive mantle or forest conditions can, at a point in time, negate the advantages stemming from the availability of suitable terrain or physical conditions. The reverse is not true for positive mantle or forest characteristics cannot offset unsuitable physical site conditions. In addition accessibility and legal factors may be further constraints which can offset both favourable physical and mantle conditions. For example where the location of a forest in respect to the regional road network imposes high time costs on visitors or where the Forestry Commission has a leasehold interest which restricts the uses it can make of a forest the potential of otherwise favourable conditions for recreation may not be realisable.

The relative suitability of physical and mantle conditions for recreational use varies within a forest, as well as between forests, because no forest can ever be completely homogeneous. To achieve a uniform crop may often be a logical aim for commercial forestry but even so no extensive area of forest will retain any high degree of uniformity for long because nature, through differences in depth and fertility of soil, in topography and exposure, in damage by windblow, etc., imposes its own diversity (Wood and Anderson, 1968). Intra-forest differences may be further compounded by rotational practices which bring about variation in the age and therefore height of trees and planting practices which introduce a variety of trees by fitting species to the differing soil and topographical conditions. Goodall (1973) has described the wide variation to be found amongst terrain and mantle characteristics within the national forest estate and Mutch (1972) has noted the importance of visual contrasts arising from differences in the size, age and variety of trees to the pleasure derived by informal recreational visitors to forests.

Besides the expected variation in the number and type of recreational activities that can take place within individual national forests it should be noted that a similar principle applies at the intra-forest level. Areas of totally planted land in which trees are grown under commercial conditions can support only a limited range of recreational activities because the close spacing of trees inhibits activities which need even relatively small open spaces. Age, and associated height, of trees appears to have a significant influence on the number and range of recreational activities possible in forested areas. For example, with very young trees recreational possibilities must be restricted because access carries a high risk of damage; with immature trees an impenetrable thicket is produced which precludes virtually all recreational activity; but as trees mature so the number of recreational activities possible increases. Where forestry operations - ranging from preparation for planting and replanting, through thinning and felling to constructional work associated with forest roads - are underway recreational use will be restricted in the short-run, although certain operations, such as the felling of an area of mature trees, will have longer term consequences. However few visitors see anything in timber production that is inherently antagonistic to their use of forests for recreation (Mutch, 1968). Moreover it is unlikely that all parts of a given forest will be simultaneously affected by such operations.

A more detailed consideration of outdoor recreation activities reveals some to be better suited to forest environments than others. If one accepts the premise that forest has a greater capacity for absorbing recreational visitor groups, along with their cars, transistor radios and other accoutrements, than other types of countryside then the use of forests for intensive, high demand recreational activities would be advantageous from society's point of view. This is so on two counts. Firstly, a forest setting may be attractive, in its own right, to recreational activity because it can screen an activity from outside interference as is the case where forest picnic places are created to give protection from the wind, privacy from neighbours and a southerly aspect for warmth. Secondly, even where a forest background is incidental to the benefits derived by participants in a recreational activity, an activity, by locating in a forest, can reduce or prevent the occurrence of externality situations normally associated with that activity; for example, the visual screening of caravan sites and the noise screening of motor sports which can be achieved in a forest environment. Forest may also serve as a complementary environment to adjacent non-forested land which is being used for a particular recreational activity. It is likely that the properties of forest environments such as visual separation and noise absorption are ones for which society will be increasingly grateful.

Recreational activities exhibit varying degrees of tolerance and flexibility regarding their ability to use a commercial forest environment. On the one hand are those activities, associated particularly with informal recreation, which show high levels of tolerance of most forest areas and which are relatively flexible in relation to the type of area used. Should access to or use of one area be denied a nearby, acceptable substitute can usually be found. The spatial extent of such recreational use in a particular forest is largely limited to an 'edge effect' because of people's fear of getting lost if they penetrate too deeply into plantations. Shooting and orienteering are examples of other flexible and tolerant activities which can take various physical and mantle conditions in their stride. Activities with a 'linear bias' to their practice, such as horseriding, rambling and certain motor sports are also relatively flexible and tolerant in respect of those parts of the forest route network they use at any given time. On the other hand activities needing a localised and permanent base, course, pitch of similar layout can also appreciate the value of forest settings as, for example, in the case of golf and field archery or they may find a forest setting to be immaterial, as with rugby club pitches or small-bore shooting ranges. Although such activities are less tolerant of variable conditions and are much more inflexible in their requirements they can be accommodated in a forest environment where recreation is regarded as the prime function and the forest layout is planned and managed accordingly. However, the opportunities for accommodating these activities in a forest, where timber production is the primary objective, are exceedingly limited.

Activities which require extensive open space, like moorland or fell walking and full-bore shooting ranges, may be possible within national forests where suitable physical conditions occur on land which is not planted to trees. Such land may be awaiting afforestation or replanting, or be retained in agricultural use, or represent land deemed unsuitable for planting on physical or amenity grounds. In the first of these instances any recreational use of the open land is likely to be short-term and in the second it will depend on the compatibility of the possible recreational and agricultural uses but, in the last case, the opportunities for recreational use may be much wider.

Since the majority of commercial forest in Britain is planted to coniferous species the preference of recreational users for broadleaf or conifer could be relevant to the suitability of commercial forest for recreational use. Whilst severe criticism of coniferous afforestation of 'wild' uplands is the order of the day where some recreational groups are involved (e.g. Ramblers' Association, 1971), so far as informal recreation is concerned Mutch (1968) found that over 90% of the respondents in his sample inquiry of forest users held the opinion that coniferous forest was suitable for recreation. Indeed, over two-thirds of them regarded conifers as more attractive than broadleaves. Since it could be suggested that the results were biased because people who disliked conifers did not visit the sample forests Mutch also related respondents' opinions to the experience of visiting the forest. No significant difference was found in attitudes to coniferous forests in the localities sampled between respondents on their first visit and those on their second and subsequent visits. Overall, however, it is likely that there exists amongst recreational organisations and the general public an alarming level of ignorance concerning the recreational qualities of mature coniferous forest.

Thus is may be proposed that forest, whether broadleaf, coniferous or mixed, has a high recreational potential and this holds true even where forest originates from State efforts to produce timber. Since conditions desired by a wide range of recreational activities are therefore likely to be found somewhere within the national forest estate the next step must be a closer examination of the supply-demand features of our national forests.

The Supply of Forest Resources for Recreation

National forests have come into being as a result of the afforestation policy pursued by the Forestry Commission since the latter's establishment in 1919, for few areas of standing timber (the Crown Woods are an exception) have passed into Forestry Commission management. If society is to be satisfied that the best overall use is being made of the nation's forests then, from a recreational point of view, not only the total supply of land in national forests but also the spatial distribution and quality of those forests needs to be considered.

The size of the Forestry Commission estate

The national target of some 2 million hectares of economically productive and well-managed woodland by the year 2000 will be achieved as a result of state afforestation, private planting and the rehabilitation of certain degenerate private woodlands. Currently there are some 2 million hectares of woodland, of which 81.5% may be classed as well-managed. As Table 1 showed national forests account for 40% of the total woodland hectarage, representing nearly 50% of the productive hectarage: that is, some 768,000 hectares, of which two-thirds represent the efforts of the post-Second World War afforestation programme. For reasons dictated by the economics of timber production, afforestation (both state and private) and replanting have favoured the use of coniferous species. Whereas 62% of the national area of woodland is devoted to coniferous species (Forestry Commission, 1974) some 94.5% of the Forestry Commission's planted area is under conifer (Forestry Commission, 1971B). This leaves under 50,000 hectares of broadleaved woodland in national forests.

The straightforward figure for the Commission's planted hectarage is misleading as to the full extent of the Commission's landholdings which total some 1.25 million hectares. Thus nearly two-fifths of the land area controlled by the Forestry Commission is not planted with trees. Of the 466,000 hectares of non-forested land owned by the Commission some 27% represents land awaiting planting, 36% is used for agriculture and grazing, 35% is unplantable for a variety of reasons, and 1% is taken up by forestry workers' holdings (Forestry Commission, 1972). The availability of this amount of non-planted or 'open' land in national forests may be of special recreational significance.

Further land is currently being acquired, by negotiation, at the rate of some 12,000 hectares per annum. The Commission's current acquisitions are concentrated in the upland areas of Britain and normally represent mountains or moorlands of low grazing value. The combined planting and replanting programme of the Commission now aimed at for the mid-1970s is of the order of 22,000 hectares per annum and represents an 8% reduction on the annual rate of the late 1960s.

The regional distribution of national forests

The distribution of forested land, on a conservancy or regional basis, is depicted in Fig. 2 along with the proportion of forest in each region which is controlled by the Forestry Commission. The figure indicates, together with Table 4, that national forests are unevenly distributed between conservancies.

TABLE 4: REGIONAL DISTRIBUTION OF FORESTED LAND AND FORESTRY COMMISSION HOLDINGS, 1973

Conservancy or Region	Forest as % of All Land Use	F.C. Holdings as % of Forested Land
North Scotland	6	51
West Scotland	16	62
East Scotland	12	39
South Scotland	12	55
North Wales	11	61
South Wales	10	61
North-west England	6	46
North-east England	7	48
East England	5	28
South-west England	7	28
South-east England	13	17

The Forestry Commission manages less than one-fifth of forested land in South-east England, which is the second most heavily wooded conservancy, but nearly two-thirds of the forest area in West Scotland which is the most heavily wooded conservancy. A rank correlation of conservancies comparing the proportion of wooded land in each conservancy with the proportion owned by the Forestry Commission reveals a positive correlation but one which is not statistically significant. Forestry Commission holdings are concentrated in upland areas, often where there is a declining population, and when conservancies are ranked in declining area of Commission woodland the order is headed by the four Scottish conservancies, followed by the two north of England conservancies and the two Welsh ones, thus leaving the basically lowland conservancies at the foot of the list. Figure 3 emphasizes the general lack of accessibility of the majority of the Commission's forests to the main population centres as a result of their upland locations. Accessible lowland forests are least numerous and least extensive amongst the Commission's holdings. Moreover, whilst such lowland forests have a key role to play in meeting recreational demand in areas where there are strong claims from land uses incompatible with recreation, the availability

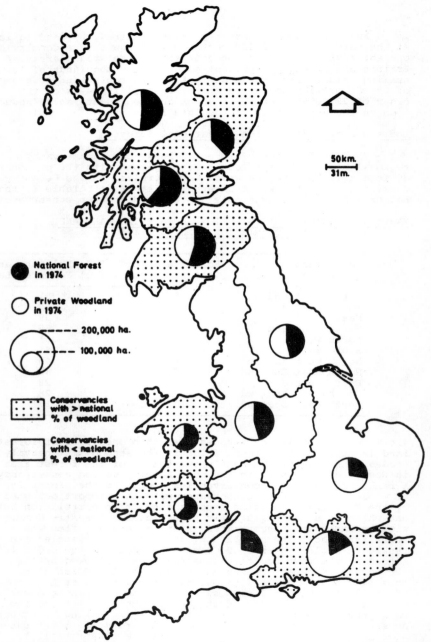

Fig. 2. Relative and absolute distribution of forest land by conservancy and national forest as a proportion of all forest land within conservancies.

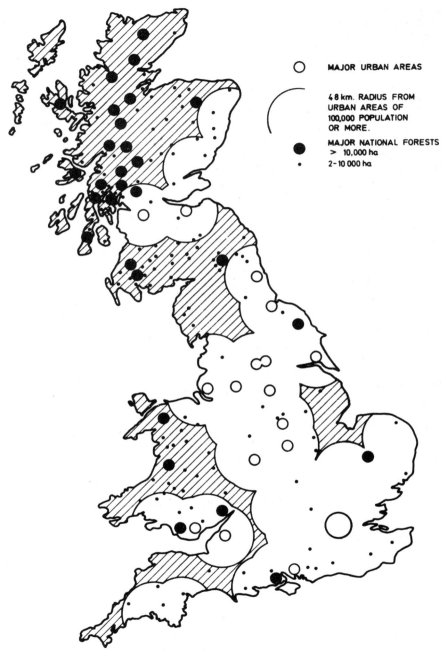

Fig. 3 The location of major national forest in relation to urban areas.
(after Grayson, et.al, 1973)

of the Commission's lowland holdings for public recreational use if often constrained by tenure conditions. Furthermore, since land acquisitions and planting are now concentrated in upland areas current afforestation will not add to the supply of lowland forest for recreation. Thus the recreational potential of new planting is also limited by its accessibility.

Some 14% of the Commission's landholdings have been designated as Forest Parks (nearly 15% if the New Forest is included). The spatial distribution of these Forest Parks reflects the overall distribution of the Commission's holdings since there are four parks in Scotland, one on the Scotland-England border, one in Wales and another on the Wales-England border. Since five of these parks are north of a Carlisle-Newcastle line their recreational potantial is only realized to the full during summer months when long-stay visitors spend annual holidays in such regions, except for any instance where winter sports can be accommodated.

Forestry Commission holdings are quite common in England and Wales in areas designated as National Parks: indeed Snowdonia Forest Park is contained within the Snowdonia National Park and Border Forest Park adjoins the Northumberland National Park along the latter's western boundary. Overall Forestry Commission holdings account for some 6% of the total area of National Parks (and only the National Trust appears as a landowner of similar scale in these areas). Nearly one-fifth of Commission land in England and Wales lies within a National Park. In practice there is considerable variation between national parks in the relative importance of the Commission as a landowner, ranging from 17% ownership of land in the Northumberland National Park, to 13% in the North York Moors, 10% in Snowdonia, 7% in the Brecon Beacons, 5% in the Lake District, down to 2% in Exmoor and Dartmoor and 1% or less in the Peak District, Yorkshire Dales and Pembrokeshire Coast parks. The Forestry Commission appears as the largest public owner of land in the Brecon Beacons, North York Moors and Snowdonia parks but in the other cases either the National Trust, water authorities or the Ministry of Defence have larger holdings (National Park Policies Review Committee, 1974). In four national parks - the Brecon Beacons, Northumberland, North York Moors and Snowdonia - well over half the forested land is owned by the Commission. Since one function of national parks is to provide areas in which outdoor recreational pursuits can be enjoyed it can be argued that the recreational use of Forestry Commission holdings in such areas should be given special consideration.

A qualitative view of national forests

The recreational quality of forest may be examined at both national and local scales. Here emphasis is laid on the former and reference made to the aggregate position as regards tree height, age, species and the like. As already noted the older the tree the taller and more attractive it is likely to be. Moreover it is not until trees are older than 25 years that their height generally exceeds 10 m and the plantation becomes penetrable to any extent following the first thinnings. Since the Forestry Commission was only established in 1919 there is little prospect of the Commission holding large areas of older trees unless they had acquired standing timber. Even in those cases there is a high probability that the timber was felled during the Second World War. It is therefore not surprising that only 2% of the Commission's forest hectarage is accounted for by trees planted before 1920 and that over three-quarters of its hectarage is represented by post-1946 plantings. Thus much of the national forest estate is of an age at which the greatest damage could result from public access as well as offering an environment which is relatively unattractive to a wide range of recreational activities. Trees exceeding 20 m are uncommon and comprise either stands taken over by the Commission and retained for amenity purposes or stands of fast-growing young conifers.

In Great Britain as a whole approximately 62% of forest is planted to coniferous species and 38% to broadleaved species (Forestry Commission, 1974), whereas the respective figures are 94% conifer and 6% broadleaf when Forestry Commission plantings are considered on their own (Forestry Commission, 1971B). This is only to be expected in view of the Commission's timber producing role and the fact that it accounts for nearly two-thirds of productive forest. Indeed, as Table 5 shows, five basic species - Sitka spruce, Scots pine, Larches, Norway spruce and Lodgepole pine - account for over 80% of the Forestry Commission's planted hectarage, compared with only 54% in the case of all national woodland. The picture of National forests is thus one of very young to semi-mature forest, dominated by a few coniferous species, with the proportion of trees planted within the last 10 years increasing northwards. Whilst the above picture is dominated by conifers it is important to note that broadleaves under Commission management also show a concentration on post-1946 planting, since some 60% of the Commission's broadleaf hectarage has been planted since that date. However, nearly one-fifth of the broadleaf hectarage in national forests was planted before 1920 (cf. 1% of coniferous hectarage).

TABLE 5: COMPARATIVE PERCENTAGES OF PLANTED HECTARAGES BY SPECIES FOR BRITISH WOODLAND AND FORESTRY COMMISSION HOLDINGS

Species	% of British Woodland	% of Forestry Commission Planted Area
Coniferous:		
Sitka spruce	20.0	35.0
Scots pine	14.0	16.0
Larches	8.5	10.0
Norway spruce	7.0	11.0
Lodgepole pine	5.0	9.0
Others	7.5	13.0
Broadleaves:		
Oak	11.0	2.0
Birch	9.0	-
Beech	3.5	2.5
Others	14.5	1.5
Total	100.0	100.0

Source: Forestry Commission (1971B, 1974).

This conifer-broadleaf picture generally holds true for national forests in England, Scotland and Wales except that for English forests the inter-war period accounts for a higher proportion of both broadleaf and coniferous planting for relatively little land has been acquired in England since 1946; for Scottish forests the effect of recent land acquisitions increases to over 80% the coniferous hectarage planted since 1946; and for Welsh forests there is a relative deficiency in respect of pre-1946 broadleaf plantings. Broadleaves, as might be expected, are most common in the national forests of Lowland Britain but the hectarage planted to broadleaves never exceeds 50% in any particular forest and, only in the case of the South-east England conservancy, was it always above 10% of each forest. In upland conservancies broadleaves rarely account for as much as 2-3% of the planted hectarage (Goodall, 1973).

Most post-1946 planting displays a greater uniformity, particularly in relation to trunk size, than pre-war planting. Within any forest variation usually derives from the fact that no national forest used less than 10 different species (and in those with a broadleaf-conifer mix over 20 species was not uncommon) and from the marked contrasts which appear when planting arrangements, as reflected in terms of tree age and height distributions, are considered. For management purposes each forest is divided into compartments ranging from a few to hundreds of hectares in extent (10 hectares usually a minimum) since they are planned on the assumption that sooner or later fire could break out. A compartment is rarely planted to a single species, although all of it is normally planted at the same time, but mixed broadleaf-conifer arrangements are generally uncommon except in Lowland Britain where, for example in the South-east England conservancy, they can account for up to 40% of a forest's compartments (Goodall, 1973). The age, and therefore height, variation amongst the trees, resulting from phased planting and rotational practices, is of particular importance in introducing variety to an area where afforestation proceeds over an extended period of time.

As noted earlier the Forestry Commission holds a large area, equivalant to nearly two-fifths of its total estate, of 'open' or non-forested land. Moreover, only a quarter of this is awaiting planting. The distribution of this open land has obvious implications for forest landscapes and for the range of possible recreational activities. In general open land not scheduled for planting is concentrated, not dispersed in small plots through a national forest and represents good agricultural land in valley bottoms or moor- and mountain-land above an 'artificial' tree line. It therefore follows that open land is not equally available throughout the national forest estate but is concentrated in upland conservancies, particularly Scottish, where land acquisition has proceeded by obtaining complete landownership units as they became available in contrast to the inter-war situation in Lowland Britain where the Forestry Commission was usually only offered land of little or no agricultural value.

Every commercial forest requires the development of a network of roads over a rotation. At the afforestation stage a low density network of 1-2 metres per hectare is sufficient for ground preparation, planting and fire protection. Additional roads are provided before the first thinnings to give a combination of road transport and off-road extraction systems which minimise the cost of moving timber from stump to mill. Currently the optimum logging road density is between 10 and 20 metres per hectare depending on terrain. Such forest roads have a formation width of 5.5 m, a carriageway width of 3 m, curves not exceeding a 45° radius, gradients of not more than 1 in 10 and are finished with a running surface of fine gravel. Over 16,900 km of such forest roads exist, although a quarter of them, built before 1950, are in need of improvement to cope with modern timber traffic (Forestry Commission, 1974). Large, established national forests therefore have road systems which offer circuit possibilities for there is a choice of route between at least two places whereas branching networks, which offer no choice of route, are more common in forests at the planting stage and are also characteristic of the watershed perimeters of upland forests. Although these roads are built to high standards in order to take heavy logging traffic they are not public rights of way and since they are single-lane roads their uncontrolled use for recreational traffic could be very disruptive to forestry operations. However, where circuit possibilities exist some one-way systems may be developed for scenic drives or made available for car rallying.

The presence of wildlife also contributes to the quality of a forest environment for recreation. Indeed, as Grant (1971) has pointed out, forest wildlife has been sadly underestimated in the past as a recreational resource for the potential in natural history study, photography, fishing and even just observing animals is immense. Conservation principles are currently practised as part of the general management of Forestry Commission woodland together with more specialized treatment for areas of specific interest. Since most conifers planted in our national forests are exotics they nearly always have poorer fauna than native broadleaf forests since only foreign tree species have been imported, not their associated wildlife. Moreover, in a commercial coniferous forest much wildlife may be discouraged by the paucity of ground vegetation and the lack of rotting and dying trees. The rapid spread of deer, which has gone hand in hand with new planting, and the increase of rarer mammals, such as pine martens and wild cats, show that some wildlife can benefit. However, forest management could do much to encourage an increased variety of wildlife (Steele, 1972).

Whilst the quality of a forest environment, in terms of the above factors, may be eminently suited to recreational use the existence of legal constraints may restrict the Forestry Commission's ability to make available and to develop the recreational potential of part of the national forest estate. Nationally about half the Commission's forests are affected by such legal constraints although in the case of over one-third of the affected forests an early opportunity to remove the constraint is associated with sporting rights which fall due for renewal by the mid-1970's (Forestry Commission, 1971C).

Measuring the Recreational Quality of a Forest Environment

The relative quality of individual national forests may be compared in terms of their recreational supply capability or potential by means of a combinatorial index based on the characteristics discussed in the first section of this chapter. Such an approach is subjective in so far as the selection of characteristics to be included in the index is concerned but selection of both topographical and mantle characteristics was related to a detailed survey of user requirements of some 120 recreational activities (inclusive of sub-variants) (Goodall and Whittow, 1973 and 1975). Objective values could then be established for each of the characteristics such as the proportion of slopes of various gradients, the proportion of broadleaf plantings, the average height of trees, the amount of open land, etc. Further subjective evaluation is then necessary to combine these elements into a single index since, for example, trees of different height and slopes of different steepness vary in their attractiveness to or desirability for recreational activities. Hence the proportions occuring in different tree height or slope categories must be weighted according to perceived importance of tree height vis-a-vis gradient and other factors.

Where the object is to decide which are the more and which are the less suitable national forests for recreation the measurement of recreational quality for any given national forest may be determined from a random sampling of characteristics within that forest. Analysis of data from a stratified sample of Forestry Commission holdings(Goodall and Whittow, 1972), weighted according to the relative importance of characteristics as discerned by recreational organizations, was used to construct an index against which the character of any forest could be measured to reveal its recreational potential. The basic form of the recreational potential index (Goodall and Whittow, 1973) is

$$P_R = P_1 + P_2 - P_3$$

where P_R is a measure of forest recreational potential on a scale from 0(low) to 100(highest),

P_1 is a measure of topographic suitability based on gradient, undulation, presence of water, amplitude of relief and elevation,

P_2 is a measure of mantle suitability based on species diversity, tree height and spacing, and presence of open land,

P_3 is a measure of accessibility/uniqueness based on intra-forest accessibility and proportion of woodland in a region.

Spearman rank correlation tests of the pairwise relationships P_R/P_1, P_R/P_2, etc. showed that P_1 and P_2 are both significant in determining the rank order of P_R and that both are positively correlated with P_R, whereas P_3, treated as an access penalty, is negatively correlated with P_R.

Scores on the P_R scale are related to the capability of forests to meet conditions specified by recreational activities and may be related to the number of possible terrestrial recreational activities in terms of a potential forest activity zone as illustrated in Fig. 4. The inflexions in the AA and BB boundaries represent thresholds on the scale: the lower is based on the increasing importance of gentler slopes and the higher on a significant proportion of mature trees. Hence a forest with a low P_R score is likely to be characterized by steep slopes, young trees and poor intra-forest accessibility and its recreational capability will be largely confined to activities which can use linear forest routes and steep slopes. Alternatively, a forest with a high P_R score has few really steep slopes, large areas of mature trees and, possibly, some open land as a result of which it can accommodate a wider range of recreational activity. Highest scores are likely to be achieved by mature lowland forests which have a fair representation of broadleaves and in which major water bodies and areas of open land are present, whilst low scores will generally be associated with young, upland forests without major water but with a very high proportion of steep slopes and which have been recently planted over a short span of time. At the intra-regional and sub-regional scales, where similar lithology may occur, it is likely that variation in P_2 scores will be the main reason for differences in recreational potential between national forests in that area. P_1 may be more important in influencing differences at the interregional scale.

The P_R score for any forest will change with time. Although P_1, the topographic element, is constant P_2, the mantle factor, is dynamic being particularly dependent upon tree height and felling practices. P_3 also changes, but to a lesser extent, as a result of management practices. Figure 5 illustrates the likely trend of the P_R score of a hypothetical forest in which planting has been concentrated into a short span of time. Recreational activity, formerly on open land awaiting planting, is restricted as soon as preparation for planting beings and access is increasingly impeded as the growing young trees interlock to form thicket. The P_R score rises quite sharply as the trees mature but will then drop when clear felling takes place. The rise and fall of the P_R score for any given forest will be a function of the phasing of its rotation.

At the level of the individual forest the indexing technique will only give a general indication of a forest's recreational potential. It will not identify the particular sites within that forest which have the greatest advantages for recreation. There are two ways of overcoming this. One possibility is to use the same characteristics as for the index above but to obtain measurements on a grid square basis to give a complete coverage of a forest and so develop a potential surface which reveals variation in recreational quality within a forest. Another possibility is that reliance is placed on the detailed local knowledge of forest managers to identify the most likely sites for recreation.

Since mantle characteristics of national forests change with time the maintenance or improvement of recreational potential in the case of any given forest is dependent upon

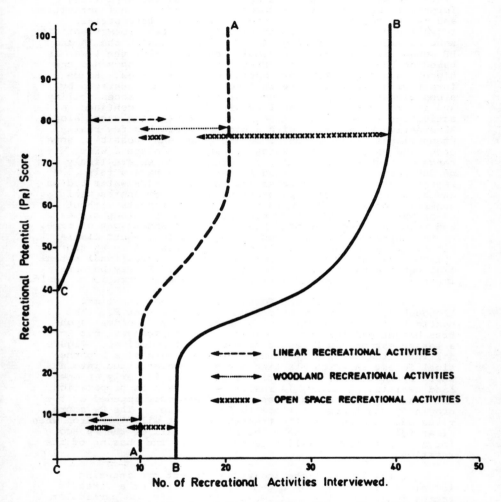

BB = max. no. of recreational activities possible within national forest estate.
AA = no. of recreational activities possible within forested area of national forest estate.
CC/BB gives no. of recreational activities possible in a given national forest.

Fig. 4 Potential Range of Forest Recreational Activity.

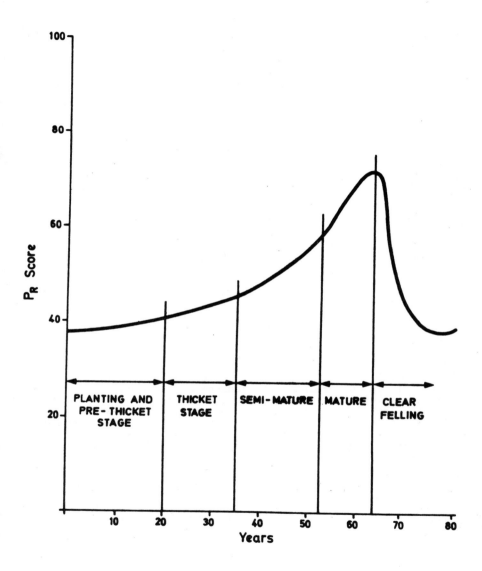

Fig. 5 Trend of P_R Score for hypothetical forest over time.

management practices. Moreover, as three-quarters of the
afforested land in national forests has been planted since
1946 the future recreational potential of national forests
is considerably higher than the present potential. This is
because the P_R scores of many national forests will rise as
the trees mature and may be enhanced further where deliberate
adjustments are made to management practices to favour
recreational ends. Current planting programmes, however,
will not have a significant recreational value until after
the turn of the century by which time the hectarage of mature
trees in national forests will probably be three times as large
as now.

The Demand for Forest Recreation

It has already been noted that few recreational activities are unique to forest environments and, also, that changes
in rural areas outside of forests are likely to reduce the
availability of other sites for recreation. Since all recreational activities typically associated with forest can usually
be carried out equally well, though with a different emphasis,
on other rural land what part of total countryside recreation
demand will focus on forests and woodland? Not only do
recreational activities adapt themselves to forest environments but it is also possible that forest environments can
be modified or adapted to suit recreational requirements.
The considerable recreational use currently being made of
national forests, particularly informal day-visitors, was
described in Chapter 2. It is not altogether clear, however,
whether this demand reflects the fact that forests have a
particular attraction for recreation, as is the case with
orienteering, or whether it is because countryside recreation
is being frustrated elsewhere and forests offer acceptable
alternative sites, as may be the case for car rallying and
for informal recreation in some areas.

From the demand side general acccessibility is an
important factor determining the recreational use made of a
given forest. For much informal recreation close substitutes
may be available within the area to be visited and a national
forest which suffers an access disadvantage relative to substitute locations is unlikely to attract many such visitors
whatever its rating on a supply capability scale. For
specialist activities inaccessibility may be less of a draw-
back, particularly where there is a paucity of substitute
locations for the activity in question. Thus the level of
recreational demand for use of a given forest is a function
of the interrelationship between accessibility and type of
recreation. The highest levels of demand will be experienced
by national forests located at the periphery of large urban
areas since large numbers of visits will be made to use the
forest as a children's playground and for exercising the family
dog, etc. Next come forests which are within an urban hinter-
land as regards easy day-tripping from the urban area: these
will attract, in particular, weekend visitors to picnic, etc.,
but will also experience general pressure during periods of
fine summer weather. However, many national forests, as
shown in Fig. 3, fall outside zones of major day-trip pressures
and levels of demand here will be much lower. The exception
is where such national forests occur in major inland holiday

areas, like National Parks, in which case they will attract a share of summer holiday visitors. The above generalizations certainly apply to day-visitor demand but the picture is less clear for specialist demands and overnight visitors. The pattern of demand in these cases is less regular, with overnight visitors concentrating where facilities are available, which is largely within the Forest Parks, whilst specialist activities are reliant upon particular conditions being available and these, as in the case of game fishing, may only occur by chance within a national forest.

Given that accessibility is a critical factor for informal day-visitor recreation the Forestry Commission's Planning and Economics Branch has attempted a gravity model approach to ascertain the level of demand at many national forests by adopting a simple trip attraction model of the form $y = d(1 - d/30)$, where y = the participation rate per 1000 car owners, d = distance in miles and 30 is an arbitrary limiting radius. In addition, 1968 levels of day-visitors were compared with a measure of recreational attractiveness of forests in each conservancy, derived by summing the product for each forest of area and a trip attraction index (assuming a linear fall-off in trip rates per thousand car owners) and weighting for the age distribution of trees. Regression of the 1968 levels of visits on this attractiveness index yielded a level of explained variation for r^2 of 0.66 (Grayson, et al., 1973).

Levels of effective demand for forest recreation may be expected to rise over time as latent demand is tapped. Latent demand comprises both deferred demand, which is non-participation in activities because facilities are not available or their existence is not known, and potential demand, which is demand that may become effective at some future date as a result of a change in the circumstances (e.g. a rise in car ownership levels) of the non-participating population (Countryside Commission, 1970). Factors such as rising incomes, car ownership and educational levels, etc., together indicate a possible doubling of participation in countryside recreation in the decade 1973-83 (Lloyd, 1972). A first reaction would therefore be to expect a doubling of recreational visits to forests and the prediction of Grayson et al (1973) that the national growth in day-visits to forests will increase by between 5 and 10% annually for the next decade or more is roughly in line with this.

Attempts have also been made to build models which predict recreation demand in the case of individual forests. In a study of the Forest of Dean nearly all of the several, different predictive techniques used confirmed that the number of day-visitors to the Forest of Dean would increase rapidly, probably up to three-fold between 1968 and 1981 (Colenutt and Sidaway, 1973). A similar rate of increase was predicted for day-visits to the New Forest (South Hampshire Plan Advisory Committee, 1969) but this is partly accounted for by the planned population growth of the region. Data from the Dean study of 1968 was used by the Forestry Commission (1971C) to establish distance factors to weight the number of cars in each 8 km (5 ml) zone up to 48 km (30 ml) for all national forests within 48 km of urban areas of population 100,000 or more or within major inland holiday areas in order to calculate the potential recreational demand for each forest and so provide some objective assessment of priorities for recreational development for day visitor purposes. The study concluded that, of the 60% of national forests accessible to large urban areas or in holiday areas half were immediately available and were likely to be particularly suitable for recreational development in view of their demand potential. This, however, presupposes that there are no differences in the drawing power or attractiveness of forests and this needs to be checked by means of the recreational potential/supply capability index.

The uncertainties which surround the prediction of a doubling of visitors to forests in a decade should not be minimised for the techniques of forecasting recreational use remain in their infancy and the projections are prone to error. Current visitors to national forests come largely from a middle-class background and the evidence to hand might suggest, according to Lloyd (1972) that increasing car ownership, which will come about as a result of increased ownership levels amongst urban working class persons, need not lead to an increase in visits to the countryside since working class people with cars do not use them to visit the countryside to the same extent as the middle-classes. Therefore, unless the urban working class changes its recreational habits, a doubling of car ownership need not lead to a doubling of forest visitors. Recreational tastes may change generally and, for example, any substitution of alternative recreational activities, such as water sports, grass skiing and special developments in country parks, may siphon off demand from forest areas. Attempts have been made, as in the Pilot National Recreation Survey (British Travel Association/Univ. of Keele, 1967), to establish latent demand for particular recreational activities. Of the activities studied in the pilot survey which take place in forests there is an indication that horse-riding and wildlife study will show the most rapid increases in participation but hiking and camping will remain the most popular activities. Whilst such studies have obvious limitations the implications of any general increase in recreational participation and any changes in demand for individual recreational activities for the planning of recreational facilities, both within and without, forests, must be borne in mind.

4. PLANNING OF FOREST RECREATION: PROBLEMS AND PRINCIPLES

Planning for Recreation: Forest versus Non-Forest Environments

The essence of recreation is choice and the attraction of forests to recreational activity is relative, being dependent upon the alternative opportunities available.

For purposes of recreational planning various classifications of countryside recreational activities may be employed but one categorisation which is particularly useful is that which distinguishes free-ranging activities from activities focused on a specific attraction or which require fixed equipment or specific site conditions (Chilterns Standing Conference, 1971). The degree of dependence of activities on scenery may be related to this classification. Amongst free-ranging activities informal or passive recreation is dominant and the three basic resources around which successful recreational use is built are topography, water and wildlife. In such cases people visit the countryside to enjoy the beauty of an area by driving, walking or riding cycles and horses through it, or just to relax, picnic and admire the view in attractive surroundings. Apart from relevant local access conditions and some informative/interpretative guidance such activity needs little formal provision of facilities. The type of land use and the standards of management of the countryside will be important factors in determining the relative attractiveness of different areas for informal recreation. In general the number of visitors will be greatest where involvement is on a short-term or day-trip basis although some weekend or longer stays can occur in connection with caravan rallies, youth hostels, scout camps, etc. Formal or organised recreational activity is numerically of much less importance amongst the free-ranging category although orienteering and motor car rallying, along with rough shooting and fishing might be appropriately treated under this head. In these instances scenery per se is of secondary importance to other factors in contributing to the satisfaction derived from the recreational experience.

Specific attractions forming the focus of a recreational outing are many and might include famous viewpoints - where scenery is obviously important - as well as stately homes, gardens, zoos and country parks where the general setting is of less importance. Such locations, even in the case of informal activity, experience concentrated recreational pressure and ancilliary facilities, alongside any special development, are essential.

Fixed equipment and specific site activities attach little value to scenery, particularly in the case of spectator sports. In a few cases such as permanent sites for camping, touring caravans and holiday chalets an element of setting is important. In other instances specific site conditions, such as those required for white water or slalom canoe racing, may occur only in areas of high quality scenery. Generally, however, the focus is the activity itself and scenery is of no importance in the case of activities such as gliding and flying, water-skiing and power-boat racing, motor cycle scrambling and car trials and all organised games. Heavy investment in immobile facilities may be necessary to provide for some of these activities, the more so if overnight accommodation is required in addition to specific facilities.

Given the extent to which activities are dependent on general scenic conditions, existence of features of long standing, specific site conditions or provision of fixed facilities, the actual sites chosen for a particular recreational activity do not have to be in a forested environment. It has already been emphasized that recreational activity is not unique to forests. However, whilst it is a matter of chance whether intensive recreational possibilities such as beaches, lakes, major rivers and stately homes occur in national forests, for many recreational activities forest environments may be amongst the alternative sites and locations to be evaluated. This evaluation cannot ignore the capacity factor and ought to consider whether society derives the greatest net benefit when recreation takes place in a forested or non-forested environment.

Landscape characterized by forest may offer certain advantages for recreational use, both from an individual and a societal point of view, in terms of the privacy and screening that can be achieved. Whilst areas fully planted to trees cannot cater for activities requiring extensive open space, such as full-bore shooting ranges or gliding and flying club airfields, they can accommodate other activities which require smaller open spaces. Some persons will choose to partake of informal recreation in a forest environment because they seek peace and quiet and view the forest as a sanctuary from modern bustle where they can feel alone (yet, in practice, may be quite close to others). For other recreational activities there may be an advantage to society in general if they are undertaken in a forest setting which screens them and so reduces or eliminates their otherwise intrusive effect on landscape. Indeed, a forest environment has a high capacity to absorb recreational activity: hundreds of people on a beach, or in a village street, or an open fell or moorland will appear as a crowd whereas in a forest setting they pass almost unnoticed.

Forest therefore has a high capacity to absorb people, one which is probably only exceeded by those cases where people are prepared to tolerate very close proximity of fellow leisure-seekers, as in the case of the more popular seaside resorts. However the capability of forest ecosystems to sustain comparatively high levels of recreational use over time must also be considered since such levels of usage may be damaging to forest environments over a long period. Whilst there are levels of usage which, if maintained, can only lead to the deterioration (and even disappearance) of the forest environment it is likely that sound management may allow higher use levels to be maintained for forest environments than for alternative environments if, for example, a comparison of the wooded and open heathland areas of the New Forest or the open fells and Grizedale Forest in the Lake District is anything to go by. However, although forest has tremendous advantages as a possible source of recreational site supply it is unlikely that the overwhelming majority of countryside recreational activities will take place in the future in forests. If for no other reason than maintenance of a degree of choice provision of further opportunities on non-forested land will be encouraged. Hence a major function of recreational planning at national and regional levels is to strike an appropriate balance between the provision of facilities and the opening up of opportunities on forested and non-forested land. Factors such as the spatial distribution of forest, the quality of forest in terms of age of trees, etc., will be relevant to the balance that is desirable at any given time in a country or region.

Multiple use - A guiding principle in the recreational use of forest?

Recognition of the increasing recreational significance of forests is merely one reflection of the versatility of forests and the complex influence they exert on total environment. Effective policies for the use of forests are based on a full understanding of the nature and relative importance of the functions forestry can perform. These functions are determined by a complex of economic, social, strategic and physical factors, including the level of development of society and the absolute and relative areas of forest in relation to population density and other forms of land use.

What is the best mix of forest products from the nation's point of view? Forests supply services as well as products. Since the time-period involved in the production of timber is an extended one the intervening period may as well be used to supply services. Hence the possibility of multiple use of forest appears as an established principle in many countries. In British national forests the production of timber is the principal function of forestry although this is complicated by complementary (or even conflicting) functions. At present it might be suggested that interaction between timber production and recreation can be ignored on a national scale because levels of recreational use of national forests as a whole are relatively limited when compared to the total recreational market. However where timber production is taken as the principal function of forestry this will place constraints on the recreational benefits realisable from forests, particularly if recreation and timber production prove to be competitive uses of forests.

To what extent are timber production and recreation compatible uses of a given forest environment? The more important timber production is judged to be, relative to recreation, the more numerous will be the constraints facing recreational use of that forest environment. The reverse reasoning also applies since sustained and intensive recreational use of a forest will place limits on the production of timber. Even so, multiple use of a given forest remains attractive because it allows management to keep open a number of options, allowing adaptation to changing market conditions.

Multiple use, however, can imply various combinations of use intensity of the activities involved and three relationships between products - joint, complementary and supplementary supply - may be relevant to the combinatorial use situation. Firstly, where products are in joint supply the amount produced of one product determines the amount produced of another product. This holds true for both main and by-products as well as for two or more products of equal status produced from the same process. Secondly, products may be in complementary supply in the sense that the production of one commodity benefits the production of another commodity, as with a leguminous crop preceeding a cereal one in a farming rotation. The relationship between product benefits may be reciprocal or non-reciprocal. Thirdly, products may be in supplementary supply in the sense that production of one product is a sideline introduced to promote fuller use of resources employed to produce the main product. Demand for resources to produce the main product exceeds that minimum threshold level necessary to justify their employment in the first place but leaves some spare capacity which can be employed at no extra marginal cost (in the case of those resources) on the supplementary product.

Elements of all three of these relationships are relevant to the multiple use of forests. In so far as the amount of afforestation for timber production in the case of national forests in Great Britain largely determines the amount of forest available for recreational use an element of joint supply, with recreation enjoying at best by-product status, is evident. A joint supply situation also exists in those instances where the recreational use of forest has direct repercussions for the output of timber. It is most likely that the development of much recreational use of our national forests to date has been in supplementary product form. Very little has therefore been demanded in the way of additional resource inputs but the fullest use of variable factors used in timber production has been promoted. This suggestion is, to some extent, confirmed by the fact that much development has been dependent upon local initiative and effort.

As complementary products timber and recreation may generate reciprocal benefits. For example, trees may offer screening advantages to recreation whilst particular recreational activities, such as deer and grey squirrel shooting, may be beneficial to timber production by reducing forest pests. Therefore, recreation as a by-product or supplementary product, and in limited cases as a complementary product, is consistent with timber production being the principal function of forestry, since levels and types of recreational use will be constrained by the timber production priority. However, where forests are under increasing pressure to accommodate more recreatinal activity the joint supply situation where both products are of equal status is more relevant. Situations could occur where recreational use is judged to be the more valuable product of forestry and it would appear desirable to free recreational use from the timber production constraint in such cases. Joint products may then be integrated or segregated according to the main objectives considered appropriate after a market appraisal in the case of any given national forest.

Further examination of joint product theory reveals that conflict may arise between the joint products as attempts are made to increase the output of one or both. Assuming recreation and timber to be joint products of the forest environment then multiple use is under pressure on two counts. First of all, timber production is being simplified in response to economic pressures which encourage a tendency towards uniformity of species and age classes, a shortening of rotation periods, etc. The implication is that the forest will be less able to accommodate multiple uses such as recreation because products become less compatible as the intensity of one or both increases. Intensification of timber production brings conflict and competition which means that recreation, as a joint product of forestry, is incompatible with high timber production use levels. Timber production forests may be likened to industrial units so that their attraction to the public will be limited, perhaps, simply to guided tours of forestry operations (Richardson, 1970). Secondly, increasing recreational demand implies conflict with timber production because of the desire, for example, to retain mature trees beyond their normal rotation age as the intensity of recreational use rises. Again, if timber production is afforded the priority, multiple use will imply a restricted, even declining, intensity of recreational use because any higher level will mean incompatibility of recreation and timber production. The extent to which management can offset such conflict is limited and management will therefore be faced with decisions concerning the allocation of forest land amongst competing activities.

The determination of priorities should not be pre-judged in favour of timber production. Detailed investigations of the relationship between timber production and recreation are, however, only in their infancy with limited studies of the physical trade-off between timber and recreational outputs, the comparative cost of timber and recreational outputs, the comparative cost of timber and recreational production in the trade-off range and the relative benefits derived from amounts of each form of output. The task is made more difficult because of problems arising in the estimation of recreational benefits. Moreover, the generation of recreational benefits, particularly from day-visitor activities, is not directly dependent on the total area of forest but reflects the fact that people do not wander far from their cars and therefore is associated with certain access points, parking areas and picnic places. Demand may be spatially directed via the provision of access, parking and picnic facilities and management may thus view the recreation-timber trade-off as being largely restricted to small parts of the forest. In the case of a typical development - a 60-space car park, adjacent picnic area and way-marked walk - some 2.4 hectares of land are likely to be affected in terms of the recreation-timber trade-off. Assuming 18,000 visits each year of one hour duration at a consumer surplus of £0.05 per hour such a site would yield an annual benefit of £900 against an annual cost (inclusive of an adjustment for timber foregone) of £410 (Grayson et al., 1973). Should it be thought desirable to retain the trees beyond their normal timber rotation age an opportunity cost of retention arises. Grayson et al (1973) estimate that, with a constant recreational benefit of £50 per hectare from the 25th year after planting and current market prices for timber, a representative crop of Sitka spruce would have its optimum felling age, 40 years, delayed by about 12 years. If annual recreational benefits of the site increased progressively with tree height the effect is to delay the optimum felling age still further to about 55 years.

The interdepartmental cost/benefit study of forestry (Treasury, 1972) recommended a general shortening of rotation periods to 40 years along with less intensive silvicultural regimes to reduce fencing, cultivation, fertilisation and weeding requirements and less demanding species planted at wider spacings. Whilst certain changes, for example wider tree spacing, may marginally increase recreational benefits the major effect would be reduction in the level of potential recreational benefit because the shortening of the rotation period would halve the proportion of mature trees in national forests. The study, however, recognised that recreation was already an important and established product of certain areas within the national forest estate and that there would be a conflict between recreational and timber needs in these cases. It therefore restricted its recommendation for shortening rotations to those parts of the national forest estate where no such conflict existed. Only limited areas would be excluded for the study estimated that, even with the broadest definition of recreational needs, not more than 10% of timber revenues need be sacrificed.

The conflict between recreational and timber values of national forests has been seen by foresters largely as a conflict between informal recreation and timber production since the specialist activities permitted have a restricted clientele and are thought to interfere little with timber production. Even then the conflict is viewed as an areally restricted one since the assumption is that only very small areas of forest will ever be used for informal recreation on any scale. Thus it would appear that recreation is to be allowed as a joint product of sufficient importance to modify timber management programmes in very limited parts of a national forest. This stops short of suggesting that the whole of a given forest may be best managed, from the national point of view, to maximise its recreational output. Any timber output would then appear as a secondary or by-product of recreational management.

This latter hesitancy may be due to forestry policy guidelines even though the long established idea, that efficient forest management depends upon policies which are stable for long periods of time, has been replaced by an acceptance of the need for frequent policy review. But while such reviews have led to a demand for versatile species and versatile forests (Johnston, 1973) and a rapidly increasing acknowledgement of the importance of environmental forestry, the production of timber remains, and is likely to remain in the forseeable future, as the major function of national forests. The basic considerations governing forest investment will therefore remain commercial but, even when full social benefits are admitted, the rate of return on investment in national forests fails to measure up to a target rate of return of 10% (Treasury, 1972). Opportunities to develop national recreational forests, as in the Netherlands and Denmark, have not been pursued and because of the timber production constraint purely recreational forests have not emerged in Britain. The multiple use of national forests in Great Britain, therefore, tends to take place within a timber production constraint which forces recreational use of forest onto a by-product, supplementary or complementary product basis and implies low levels of recreational use intensity. True joint product status which allows recreation to emerge as the major, or even sole, product as the demand for recreation rises relative to that for timber is applied to only very limited parts of certain national forests.

The Compatibility of Recreational Activities in a Forest Environment

Recreation comprises a heterogeneous set of activities and the use of a forest environment by two or more recreational activities may generate conflict between those activities, leading to a reduction in recreational benefits for one or more of the activities involved. Compatibility of the various recreational uses of a forest is therefore a further dimension of the multiple use concept. Management decisions are broadened in so far as they have to consider possible intra-recreational conflicts as well as the recreation/timber trade-off. Intra-recreational compatibilities in a forest setting involve the balancing of space and time elements with use intensity levels.

From the spatial point of view the compatibility of recreational activities involves not only the possibility of two or more activities being able to share the use of a given site at the same time but also their ability to use adjacent sites in a mosaic of different recreational activities. Since recreational activities can have different requirements in respect of the site characteristics desired, competition for the use of a given forest site will be limited to those recreational activities whose requirements are most closely satisfied by the site's characteristics. Hence, recreational activities demanding extensive areas of open space are likely to be excluded from using forest environments because site conditions are unsatisfactory as regards the mantle element. Where recreational activities have widely differing site requirements the spatial pattern of locations used must reflect the availability of required conditions. Only where activities have similar requirements in respect of a forest environment, as in the case of rambling and hacking, does the possibility of sharing the use of a site arise and compatibility of the activities in question will depend upon the nature of the recreational activities involved, the way in which they are practised and the nature of the site or resource.

Where a recreational activity generates excessive noise (as with motor sports), introduces an element of danger (shooting), or merely involves visual intrusion (caravan sites) varying degrees of conflict or incompatibility arise in respect of the use of a site, and adjacent sites, by other activities. In some cases forest environments may increase the degree of compatibility of recreational uses, particularly of adjacent sites, where screening effects overcome visual intrusion and noise absorption properties mitigate noise penetration. In other cases, however, the degree of compatibility may be lowered in a forest setting because restricted visibility may increase the probability of accidents where an element of danger is involved in the performance of an activity.

As with the recreation-timber case, major problems of incompatibility arise as the use intensity level of one or more recreational activities using a given area of forest increases. This holds true even amongst informal or passive recreation. Incompatibility is particularly commonplace for 'linear' activities which use the roads, tracks, rides and paths through the forest. For example, as the intensity of use for horse-riding increases so the use of the same routes by walkers becomes increasingly difficult. The level of intensity at which incompatibility first becomes significant reflects both the way in which the activity is practised and the perceptual attitudes of the participants. Thus, where a competitive event is being staged any interference arising from the presence of participants in another activity would promote incompatibility even at very low use intensity levels. Competitive events, therefore, normally require the exclusive use of an area for the duration of the event. Exclusive use is also likely where permanent facilities are established for an activity. Even in the case of informal or casual participation much will depend on the skill of the persons involved in an activity, for example, the ability of a casual rider to control his horse when meeting walkers on a narrow forest track or when passing through an area used by picnickers. Perceptual attitudes can obviously differ between participants in a given activity but it is likely that even greater variations exist between participants in different activities. Moreover the use levels at which perception of incompatibility becomes apparent need not be reciprocal between two recreational activities. The occasional horse-rider passing through a picnic place can add interest but a continual stream of riders would be inconvenient and annoying. As soon as one activity perceives an incompatible situation arising from increasing use intensity management must consider the spatial separation of activities. Perceptual attitudes will also have a bearing on the maximum use intensity of a given recreational activity and perceptual capacity may be higher per unit area of forest, than in the case of non-forested areas, because of the screening effect.

Introduction of the time dimension extends considerably the range of compatible situations. Two or more recreational activities, which would be obviously incompatible if attempting to use a given forest site at the same time, may in fact require the use of that site at different times. For example, if a stretch of forest road has characteristics which make it suitable for a rally stage or a scenic drive, then the two activities are not incompatible since car rallying is a winter activity and pleasure driving a summer one. Similarly roe deer shooting can take place in a forest in the early morning before informal recreational visitors arrive. It is possible, however, that use of a site by one activity has a deleterious effect from the point of view of another activity. For instance motor sports can so churn up an area that it is unsuitable for archery or camping or dangerous for horse-riding on a subsequent occasion.

In general the more specialised a recreational activity, the more specific its site requirements, the greater its need for permanent facilities, the more frequently participation is on a competitive basis, the greater the danger element involved, and the higher its sustained use intensity level, then the more likely that recreational activity is to demand the exclusive use of a site at any given time and even over time. Since few, if any, forest recreational activities satisfy all these criteria the occasions on which a given activity demands exclusive use of an area of forest hardly ever arise and, therefore, multiple use by two or more recreational activities, as well as by recreation and timber production, is feasible. It would appear that the informal recreational activities - rambling, picnicking, horseriding, cycling and natural history pursuits, along with overnight activities such as camping and touring caravanning - form a group of activities with a high degree of intra-group compatibility in terms of use of forest environments (Goodall and Whittow, 1975). Even so incompatibilities may occur as use intensity increases, in which case conflicts will be dealt with by means of treatment similar to that given to the less compatible group of forest recreational activities (which includes most of the specialised activities discussed earlier), namely, spatial separation. It is important to note that the informal recreational group is more likely to conflict with timber production than the specialised activities currently permitted in national forests because the recreational benefits derived from the former are more dependent upon the landscape/scenery factor.

Recognition of degrees of compatibility or incompatibility are essential for management strategies designed to maximise multiple use, particularly intra-recreational, of forest environments. Where incompatibilities exist between recreational activities (or between timber production and recreational use) the solution to the conflicts can only be found by programmed separation. This may involve, in extreme cases, the prohibition of one or more activities from forest environments (although it must then be recognised that the reservation of exclusive rights for such activities in alternative, non-forested locations is then essential). More likely is the adoption of space zoning which brings about the physical separation of incompatible activities within a forest but this requires care in designating the means of access to assigned areas. Where only limited areas suitable for recreational use exist in a forest and the pressure of demand is severe, timetabling of the area's use (or time zoning) is the only way to lessen the possibility of conflict. Management is thus able to exercise some measure of choice and possesses certain strategies which allow it to promote some degree of multiple use of a forest environment.

Some Principles for the Recreational Use of Forests

The role that forests, and national forests in particular, can play in the general provision for outdoor recreation may now be summarised. Forest is not equally suitable for all outdoor recreation: for example, its ability to cater for water-based or extensive open area activities is a matter of chance. However, a wide range of recreational activity, especially of an informal nature, exists for which forest provides a desirable or essential environment or represents a setting to which an activity can adapt if no better alternative is available. Certain principles may apply to the selection of activities from within this range which are to be encouraged within the national forest estate. An obvious one is to give preference to those recreational activities which do not conflict with timber production (and other objectives of forestry policy such as conservation). This would, of course, be restrictive in terms of both the types and amount of recreation possible, since recreation would be treated as a by-, supplementary or at best, complementary product. Such a course of action, although involving no sacrifice of timber production, would generate the lowest level of aggregate recreational benefit from national forests.

A second principle might be to encourage those activities generating the highest recreational benefits from the use of forest environments. From the community viewpoint, adaptable recreational activities, which generate undesirable externality effects on non-forested land, should be steered to forest environments where those externality effects would be minimised or eliminated by forest screening properties. This would further maximise net social benefit, allowing for possible sacrifice of timber revenues. For example, the use of forest roads for car rallying may be preferable to the use of public roads. High recreational benefits may be derived from forests forming vehicle-free zones or havens of peace and quiet. The interest of the general public in wildlife (for its grace, beauty and novelty) is also a major recreational use to which forest can make a valuable contribution.

Other principles might involve any priority to be given to activities on the basis of strength of demand since aggregate recreational benefit is likely to be raised where provision emphasizes opportunities for the largest number of participants. Recreational activities make differing demands for investment in facilities. In general, it may be suggested that heavy investment in facilities for a particular recreational activity should not be considered an essential function of recreation in national forests unless it can be shown to be a collective good which cannot be provided on any private forest estate. Any positive preference for or restriction on certain recreational activities permitted in national forests should be justified in terms of net social benefit. Exercise of such choice would, currently, tend to favour mass-participation activities involving relatively inexpensive facility provision but, overall, some sacrifice of timber production.

Having selected appropriate recreational activities forest management must then decide on the scale of provision to be attempted. Measures of supply capability or recreational capacity are necessary since, as already noted in the case of multiple use, the extent to which an activity can be accommodated (assuming compatibility) depends upon the use intensity of other activities. In practice limits to recreational use have probably been decided by trial and error because of the difficulties inherent in any objective assessment of carrying capacity at each site. Levels of recreational use which lead to a degradation of the forest environment, that is, beyond which the forest loses its powers of recovery, should be avoided. This view of carrying capacity has an ecological emphasis, vital in the long-run, but there are also economic, perceptual and physical aspects to be considered (Countryside Commission, 1970).

Recreational carrying capacity or supply capability may vary between forests because of differences in the physical character of forest and in the non-recreational demands made upon it, the efficiency of forest management, the number of recreational activities permitted and their organisation, the level of investment in facilities, institutional factors and externality situations. Physical characteristics partly determine a forest's ability to withstand recreational use without deterioration - for example, some tree species are more susceptible to soil compaction than others and cannot withstand sustained pressure over many years. The more important other functions are judged to be in a multiple use situation the lower, as a consequence, recreational capacity is likely to be. Considering Dutch standards there would appear to be a considerable difference between the recreational capacities of production forests and recreational forests: some 3 to 10 persons per hectare compared to 10 to 20 persons per hectare respectively (Sidaway, 1974). In the case of open land within national forests the use made of that land, often agricultural, may limit even more than forestry the recreational possibilities. Efficient management, by judicious use of zoning arrangements, effective signposting and parking provision, may well be able to adjust ceiling use levels upwards. Since recreational activities differ in the intensity with which they use land the greater the unit area required per participant the smaller the number of people that may be accommodated in any forest. Where recreational activity is organised on a club basis discipline becomes a club responsibility and the control exercised by clubs can do much to ensure the fullest use of available resources. Management decisions to invest in facilities - whether general, such as more parking spaces, or specific, such as breeding pheasants for shooting - have obvious implications for capacity. Institutional factors, arising from legal constraints where the national forest represents a leasehold interest, and any negative externality effects following from the introduction of recreation into a forest environment serve to restrict capacity.

The amount of use is also related to the principle of whether or not to charge for the recreational use of national forests. Pricing would allow supply to be rationed but problems of collection of charges, particularly in the case of much informal recreation which uses many small, dispersed sites along public roads, means that many opportunities will continue to be available free of charge. Entry charges are, however, feasible for large, concentrated facilities, for specialised activities controlled by licences and for overnight accommodation where there is a site warden.

A further principle relates to the distribution of recreation within the national forest estate. Where, for example, recreation has full joint product status then specialisation of function between national forests would be a realistic policy with certain forests being assigned a specific recreational role. The remainder would have a basic timber production function with subsidiary conservation, water-catchment and even by-product recreational roles. Forests assigned a primary recreational role, particularly for informal recreation, must be accessible to large numbers of people, basically day-trippers, and must be within easy driving distances of large cities. Locational constraints will therefore restrict the number of national forests likely to have a primary recreation function. Private woodland is generally better located in respect of populous areas and may well complement national forests as a recreational resource base (Duffield and Owen, 1972). Even where a national forest is locationally suitable there may be severe limitations on its current recreational potential because a large proportion of its trees are less than twenty-five years of age. Such forests have a future potential which may be called upon to increase supply at a later date in response to a continually rising demand. Choice of national forests for recreational specialisation thus depends on both demand and supply factors. Specialisation of function between national forests is thus a logical extension of joint product theory where relative demand at the local scale favours one of the joint products.

A similar principle of specialisation applies to the distribution of recreational activity within a national forest and to the use of its route network. Such intra-forest specialisation will reflect the strength of competing demands and the compatibility of activities and will manifest itself in the use of careful zoning and timing arrangements to accommodate varied and conflicting activities. Zoning and timing arrangements will be related to recreational capacity and to methods of regulation of recreational use by licence, permit, club organisation, etc. The character of the intra-forest mantle is not uniform, largely because the age of trees varies due to rotational practices but also as a result of species variation where different trees are fitted to local soils. Specialisation of recreational use within any forest should recognise this variation and to this end a forest may be sub-divided into a number of environmental areas reflecting recreational suitability (Goodall and Whittow, 1973). Such environmental areas depend largely on tree height or age since the more mature the trees the greater the recreational potential but they also allow for categories of open land in national forests. Where a national forest has been developed basically for timber production but is not assigned a purely recreational function it inherits from its timber production days a variation in recreational potential at the intra-forest level. Thus the forest environmental area concept is of use in planning or zoning the recreational use of land within that forest as well as in matching by-product recreational activity to the parts of a timber production forest.

Therefore, where part of the national forest estate is to be managed primarily for recreational purposes certain silvicultural practices will need to be modified. The recreational attractiveness of forest responds to management in both large and small-scale aspects. Forest to be used for recreation, but developed originally for timber production, needs diversifying. This entails longer rotations where the whole of a national forest is concerned or retaining some areas beyond normal rotation age where only part is affected; the use of natural regeneration; cutting to maintain views and contrasts; and planting to introduce additional species (including, for example, ones which add colour in autumn). Emphasis is therefore laid on appearance and conservation of species. Many of these adjustments are most easily achieved at second and subsequent rotations (Zehetmayr, 1971), a point already reached in many lowland forests with a high recreational potential from both the demand and supply points of view.

5. A RECREATIONAL STRATEGY FOR NATIONAL FORESTS

Resource Availability and Recreational Strategy

The recent government review of forestry policy (Ministry of Agriculture, 1972) acknowledged that more emphasis should be given to realising the recreational potential of national forests. Whilst recreational needs might be a factor considered in managing existing holdings, including replanting programmes, it would appear that future recreational possibilities would not be a factor influencing the selective acquisition of land for new planting. The latter may be worthy of reconsideration in view of the fact that, of Great Britain's low forest hectarage per capita, only a small proportion is devoted to recreational use and that where forest is in private ownership it is not generally available for public recreational use. In certain circumstances recreational possibilities could be a locational factor of importance in afforestation. Current policy, however, seeks to further recreational development of national forests within a multi-objective framework and accepts constraints on recreation imposed not only by other objectives, such as timber production, but also by the locations of existing national forests. It contains the belief that satisfactory recreational conditions can be provided in evergreen forests originally planted for timber production, except in the case of short-cycle uniform softwoods for pulping which are felled at an early age.

Whilst further encouragement of recreational use is acknowledged policy the strategy - both in terms of spatial distribution of facilities and types of recreation encouraged - remains that of providing minimal investment facilities for informal recreational activity particularly where accessible to large numbers of people. Conservancy or regional recreational plans are currently being prepared within these guidelines, after which priority evaluation will precede the formulation of tactical or local plans to implement recreational development in any particular national forest. In the discussion which follows below, the policy recommendation that the recreational potential of national forests should be developed is accepted and attention is focused on the strategic implications, that is, the structural framework within which recreational development of national forests is to take place. In discussing such a strategy it is admitted that tactical issues may arise in implementing the strategy in any particular situation but these issues will not be pursued since they are best dealt with on individual merit and need not detract from the preferred strategy.

A Strategic Spatial Framework for National Forest Recreation

Variety is a desirable feature of recreation provision and, in the national context, forest opportunities are essential alternatives to other facilities. Demand for recreation represents a continuum in which urban and rural categories are not always easily distinguishable although forests will be associated with the pursuit of those outdoor recreational activities considered largely rural in terms of their locational requirements. Recognition of differing intensities of recreational land use and of the differing dependencies of recreational benefits on scenery leads to a three-fold classification of recreational areas based on the relationship of user-resource characteristics and accessibility: namely, user-oriented, intermediate, and resource-based (Knetsch and Clawson, 1967).

User-oriented areas, usually of restricted spatial extent, are close to users' homes, used for short periods on any visit and, because of the heavy use to which they are subjected the quality of natural resources is unimportant and a great deal of alteration is acceptable and necessary. In addition to urban parks, golf courses, sports fields and stadia, and even urban fringe country parks fall into this category. Intermediate areas comprise recreational resources catering mainly for day-tripping and, perhaps, weekending. The extent of the zone used is dependent upon the quality of the road system since the private car is all-important for recreational use in this case. Positive preferences are shown, where informal recreation is involved, for areas with high quality scenery, water and/or forest and for available facilities in the case of specialist activities such as sailing or motor-cycle scrambling. The degree of manipulation from the natural state is less and acceptance of quasi-urban features lower. It is probable that access conditions in a small island like Great Britain places most of the country in this category. Resource-based recreation is dependent on extensive areas of outstanding natural resources, as in the national parks of North America, and because of remoteness attracts long-stay vacationers. Environment is all-important and the impact of recreation is minimized. For Great Britain probably only the remoter parts of the Scottish Highlands could qualify under this category. This classification provides a useful spatial framework, when associated with access from conurbations, within which to discuss a recreational strategy for national forests.

With the continued rising, relative and absolute cost of private mobility, accessibility will become increasingly important in the next decade and the role of outdoor rural recreation may be reappraised with a consequent narrowing of the extent of the intermediate zone. Greater emphasis may therefore be given to the provision of recreational facilities in and around cities. For example, the Dutch suggest a hierarchy of recreation areas of increasing size and specialisation as one moves away from the residential neighbourhood, but recommend major investment in parks more than 10 km from large population centres is only justified where other objectives, say landscape conservation, are involved (Sidaway, 1974). Such ideas are consistent with the three-fold spatial framework outlined above but recognise a narrower intermediate zone. Within the intermediate zone conflicts arise with regard to the acceptance of intensive, urban-user oriented recreation into lowland agricultural areas and a clear-cut subdivision of land use is required.

In terms of a spatial strategy for forest recreational development three zones therefore suggest themselves:-

(i) Urban or urban fringe forests which cater for user-oriented activity, from walking the dog to adventure playgrounds and amusement parks. Such European examples, probably unrivalled in Great Britain, as Het Amsterdamse Bos in Amsterdam, the Marseilisborg Forest in Aarhus and the Jaegersborg Forest in Copenhagen indicate the possibilities of forest settings for such activity. In a more specialised vein golf courses incorporated into wooded areas are a common urban fringe land use. It is in this type of forest that many of Richardson's suggestions (1970) could best be accommodated.

(ii) Lowland forests providing opportunities for both informal and specialised recreational activity ranging from pleasure driving to car trials, rambling to orienteering, natural history study to shooting, picnicking to camping, etc. In these cases a forest environment may be preferred but national forests could be in competition with innovative private forest owners in attempting to meet some recreational demands.

(iii) Highland forests which, although largely the efforts of afforestation, are the nearest thing to resource-based recreation areas. Here the level of recreational activity is low and of an informal nature, except where special features, such as an opportunity for winter sports, exist. Extensive areas of open moorland are normally a feature of such forests.

Viewed in relation to this classification most national forests fall in the Highland category which severely limits their recreational potential, in amount and type, and development on any scale is dependent upon the provision of accommodation in the form, at least, of camping and touring caravan sites or log-cabins. There is no shortage in supply of national forests of this type, although some are currently constrained by a high proportion of young trees. Moreover, this is also the category in which the greatest increase in future supply will occur as a result of current afforestation. However, a not negligible hectarage of the national forest estate does occur in the Lowland category and here recreational capacity of existing resources could be considerably increased for both informal activities and specialised uses by means of zoning and timing arrangements and certainly by the development of Forest Country Parks. National forests with urban fringe locations are very few in number and none has been developed as an urban forest park on the lines of European counterparts or even to the same extent as Epping Forest. There is a lack of forest resources in this category and public bodies would have to take the initiative if recreational facilities were to be provided in association with 'green-urban' land (Fairbrother, 1970) around our cities.

The Spatial Implications of the Overall Strategy
in Depth

(i) Highland Forests

Additional recreational resources with a forest base
need to be developed as part of overall recreation policy -
in particular 'green-star' areas as user-oriented honeypots
at the urban frings and Forest Country Parks in the intermediate zone. Further Forest Parks in the Highland zone may
be desirable to bring about a better spatial balance than
exists at present in the availability of resource-based
opportunities. For example, Forest Parks could be created
in Central Wales (based on Hafren, Rheidol and Ystwyth
forests) and in South Wales (based on national forests in
the Brecon Beacons). In the south-west peninsula, where a
forest park might seem highly desirable, none of the current
national forest holdings are of sufficient size to warrant
such a development. Further forest park developments in
Scotland would be possible in Corrour, Glen Garry or Leanachan
forests and, depending on some improvement to access, in
Borgie, Shin and Torrachilty forests to give but a few
examples. A Forest Park based on Thetford Chase might also
have considerable potential since the New Forest example
shows that not all long-stay holidays are associated with
upland areas or seaside resorts. Since new planting is
currently concentrated in the Highland zone further forest
parks could be created, as part of Forest Development Areas
(Matthews, 1972) incorporating recreational objectives, if
the need arose, for example, if Scotland was to become a
major Common Market holiday area. Too much development in
the remote north-west would not be desirable and alternative
locations for parks might be sought in the south and eastern
parts of the Highlands, as at Huntly and Rannoch forests,
or even in the Southern Uplands at Glentress and Elibank.

Purpose-built second homes might be introduced into
other Highland forests although such development on any
scale should probably await the formulation of a coherent
national policy on second homes. Camping and touring caravan
sites could be introduced to upland national forests in
regions currently lacking such facilities, for instance,
Central Wales or the North York Moors. Environmental education or field studies centres might prove feasible in some
forests. However, overall access limitations, coupled with
climatic factors, constrain even resource-based recreational
use of many upland forests.

The role of forest in National Parks is deserving of
particular attention in the light of the National Park
Policies Review Committee report (1974) which offers a four-
fold classification of areas within national parks according
to the qualities and capacities of each area. In nature
conservation areas all access is to be discouraged; in wild
and relatively remote areas valued for their scenery and
wildlife car access will be severely limited; on good farm-
land or productive woodland access would be via limited
rights of way and any facilities would be self-contained;

but in intensive recreation areas large-scale facilities, including car parking, would be made available to absorb most people. Production woodlands may not always be compatible with the other aims of National Park authorities, in which case consideration might be given to national forests assuming a greater recreational role in such areas. A policy of generally opening forests to visitors on foot could considerably increase the recreational capacity of woodland in the Committee's third category. It might therefore be argued that the recreational function of national forests in National Parks (and, for that matter, in Areas of Outstanding Natural Beauty) be accorded higher priority and developed in consultation with National Park planning authorities where this will relieve pressure on the open morrland areas of the parks.

(ii) Lowland Forests

Lowland forests are in a better position to help satisfy the day-visitor recreation demand but a question arises as to the nature of their contribution. Should the Forestry Commission take the initiative in converting a number of lowland national forests to Forest Country Parks? This would have the effect of expanding the range of choice of recreationists in respect of the settings of country parks. A number of such parks are desirable. Since such developments need to be carefully integrated with area recreation strategies forest country parks are perhaps best developed on a co-operative basis with local authorities, as at Butser Hill. The possibilities are then immense and developments comparable to the sculpture park of the Hoge Veluwe in the Netherlands would not be beyond the bounds of imagination. It should be emphasized, however, that forest is but one setting attractive to day-trippers and a range of alternative environments should also be made available.

Not all lowland national forests would therefore be given over to single-purpose intensive recreational use. These other forests would be used, largely, to cater for small-scale informal recreation by means of small car parks, picnic places and forest trails in a multi-purpose forest setting. In addition to existing scenic forest drives further drives might be investigated in forests such as Sherwood (Nottinghamshire County Council, 1972), Chiltern, Bedgebury, Savernake and Wareham. Consideration might also be given to the diversification of recreational activities with more encouragement being given to the use of lowland national forests for specialised activities, such as motor sports, which wish to use forest environments. This would still leave considerable areas of lowland national forest for those who seek peace and quiet.

With both the purely recreational function and the multiple use situations, time and space zoning would be necessary to gain the highest recreational benefit since incompatibility of recreational activities must be allowed for as well as fitting recreation to forestry operations. Conservancy recreation plans, particularly in lowland regions, must therefore reflect inter-conservancy differences in demand for recreation and an awareness of supply opportunities in national forests relative to those on private forest land and on non-forested land in general. Hence the need to integrate such plans with regional recreation structure plans.

(iii) Urban Forests

National forests can make only a very limited contribution to the supply of user-oriented, urban fringe recreational facilities since, on the one hand, there are few forests in appropriate locations and, on the other hand, many of the recreational demands are for the intensive use of open space. Although there is no reason why sports stadia and similar developments cannot be situated in forest environments it is unlikely that this will take place unless forests exist in suitable locations. A revolution in forestry policy would have to take place before new planting, as Fairbrother (1970) sees desirable, was undertaken at the urban fringe even though extensive areas of woodland would not be necessary to create a convincing forest environment. It is therefore unlikely that opportunities exist to develop urban forest parks on the Scandinavian model from amongst existing national forests. Instead such forests will be intensively used for local pursuits such as walking, riding, exercising dogs and adventure playgrounds.

Thus, given the current spatial distribution of national forests, the highland location of the majority of national forests restricts the nature and amount of recreational use that can be made of them. This probably enhances the case for some lowland forests to be devoted to single-purpose recreational use as Country Parks and for others to have their recreational use extended wherever feasible (since it would be a generation or more before any specific recreational afforestation in lowland areas could make any contribution to increasing supply). It is therefore important that the recreational potential of lowland forests be appreciated and steps taken for its realisation before any widespread clear felling takes place.

Conclusion

Whilst admitting the desirability of more positive management of national forests to enhance public recreation, visual amenity and conservation values, it is unfortunate that the recreational efforts of the Forestry Commission to date have attracted less attention than those of afforestation. Much has been achieved within policy constraints in the field of recreation and innovative behaviour has not been lacking. For example, Forest Parks derive from similar principles to those behind the creation of National Parks in Great Britain but pre-date the latter by a decade. Recent countryside legislation has given added powers to the Forestry Commission to provide for recreation and tourism on its holdings. Rates of current recreational activity not only reflect levels of demand but also of supply and the creation of more opportunities would bring many more participants. The role of the Forestry Commission in developing opportunities for recreation in national forests, when integrated with local and regional planning, should be seen in terms of broadening the recreational environments from which the public can choose.

National forests are particularly valuable where private forestry is slow to respond or restricts participants, as with the Countryside Club.

It is the recreational potential of existing national forests which must be developed. However, because of other demands made on those forests, the level of recreational use which maximizes recreational benefit may not be the level of use which maximizes net social benefit because of other products or opportunities foregone. There may well be in instances where the value of the alternative uses to the nation exceeds the value of recreational use, in which case recreational use levels will be dictated by complementary, supplementary or by-product possibilities. Thus the remote upland location of many national forests, in restricting the intensity of recreational use, will allow a basic multiple use approach to management in which recreation is integrated, as appropriate, with other functions and only assumes a primary importance in Forest Parks. However, for many lowland forests, which account for only 15% of planted hectarage, recreation could well become the single or dominant use. In some cases the trees are too young to have a current recreational attraction or the forested blocks are too small and scattered to be of major recreational value.

Overall, it is therefore unlikely that more than 10% of the planted hectarage of national forests could be beneficially turned over to specialist recreational use now or in the near future. In a sense spatial specialisation of function between national forests is a variation on the multiple use theme and one which demands acceptance of the principle that lowland forests should have a primary recreation function. It is to be hoped that the management of national forests in lowland areas will view the provision of user-oriented, intensive recreational opportunities at the urban fringe or within day-trip range as a widening responsibility and that this challenge will be met as effectively as resource-based recreation is catered for in upland national forests and forest parks.

REFERENCES

British Travel Association/University of Keele (1967). The pilot national recreation survey: report no.1, London: BTA.

Chilterns Standing Conference (1971). A plan for the Chilterns, Aylesbury: Chilterns Standing Conference.

Colenutt, R.J. and Sidaway, R.M. (1973). Forest of Dean day visitor survey, Forestry Commission Bulletin 46, London: HMSO.

Coppock, J.T. (1968). Changes in rural land use in Great Britain, pp.111-125 in Land use and resources: studies in applied geography, London: Institute of British Geographers.

Countryside Commission (1970). The demand for outdoor recreation in the country, Report of a seminar held in London on 15 January, 1970, London: Countryside Commission.

Duffield, B. and Owen, M.L. (1972). Forestry policy: Recreation and amenity considerations, Tourism and Recreation Research Unit Report No. 8, Geog.Dept. Univ. of Edinburgh.

Edlin, H.L. (1969A) Forest parks, Forestry Commission Booklet 6, London: HMSO.

Edlin, H.L. (1969B). Timber! Your growing investment, Forestry Commission Booklet 23, London: HMSO.

Fairbrother, N. (1970). New lives, new landscapes, London: Architectural Press.

Forestry Commission (1943). Post-war forest policy, Cmnd. 6447, London: HMSO.

Forestry Commission (1970). Day visits to Forestry Commission land: report of the 1968 survey, Unpublished.

Forestry Commission (1971A). Fifty-first annual report and accounts of the Forestry Commission, 1970-71, London; HMSO.

Forestry Commission (1971B). Analysis of growing stock in Forestry Commission and private woodlands, Unpublished.

Forestry Commission (1971C). Recreation planning for day visitors, Unpublished.

Forestry Commission (1972). Fifty-second annual report and accounts of the Forestry Commission, 1971-72, London: HMSO.

Forestry Commission (1973). Report on forest research, 1973, London: HMSO.

Forestry Commission (1974). British forestry, London: HMSO.

Forestry Commission (1975). *Fifty-fourth annual report and accounts of the Forestry Commission, 1973-74*, London: HMSO.

Goodall, B. (1973). The composition of forest landscapes, *Landscape Research News*, 1(5), Summer.

Goodall, B. and Whittow, J.B. (1972). The recreational potential of Forestry Commission holdings, *Report on forest research, 1972*, London: HMSO.

Goodall, B. and Whittow, J.B. (1973). *The recreational potential of Forestry Commission holdings: a report to the Forestry Commission*, Reading: Dept. of Geography, Univ. of Reading.

Goodall, B. and Whittow, J.B. (1975). *Recreation requirements and forest opportunities*, Geographical Paper No. 37, Dept. of Geography, Univ. of Reading.

G.P. (1973). Rally review: the RAC rally of Great Britain, *Motor Sport*, January.

Grant, W. (1970). Trees are for people, *Forestry Commission Research and Development Paper 83*, London: Forestry Commission.

Grayson, A.J., Sidaway, R.M. and Thompson, F.P. (1973). Some aspects of recreational planning in the Forestry Commission, *Forestry Commission Research and Development Paper 95*, London: Forestry Commission.

Johnston, D.R. (1972). Formulation and implementation of forest policy, *Forestry Commission Research and Development Paper 89*, London: Forestry Commission.

Knetsch, J. and Clawson, M. (1967). *Economics of outdoor recreation*, Baltimore: Johns Hopkins for Resources for the Future.

Lloyd, R.J. (1972). The demand for forest recreation, pp.93-108 in R.C. Steele (Ed.) *Lowland forestry and wildlife conservation, Monks Wood Experimental Station symposium No. 6*, Monks Wood: Nature Conservancy.

Matthews, J.D., Philip, M.S. and Cumming, D.G. (1972). Forestry and the forest industries, pp.25-49 in J. Ashton and W.H. Long (Eds.) *The remoter rural areas in Britain*, Edinburgh: Oliver and Boyd.

Ministry of Agriculture (1972). *Forestry policy*, London: HMSO.

Mutch, W.E.S. (1968). Public recreation in national forests: a factual survey, *Forestry Commission Booklet 21*, London: HMSO.

Mutch, W.E.S. (1972). The recreational use of conifer woodland, pp.74-76 in *Proceedings of the 3rd Conifer Conference, 5-8th October, 1970*, London: Royal Horticultural Society.

National Park Policies Review Committee (1974). *Report of the National Park Policies Review Committee* (the Sandford Report), London: HMSO.

Nottinghamshire County Council (1972). *The future of Sherwood Forest*, Nottingham: Notts County Planning Dept.

Ramblers' Association (1971). Forestry: time to rethink, *Brief for the Countryside,* **3**, London: Ramblers' Association.

Richardson, S.D. (1970). The end of forestry in Great Britain, *Advancement of Science*, **27**(132), Dec., 153-163.

Sidaway, R.M. (1974). Organisation of outdoor recreation research and planning in the Netherlands, *Forestry Commission Research and Development Paper 107*, London: Forestry Commission.

South Hampshire Plan Advisory Committee (1969). *Study report no. D5: recreation*, Winchester: Hants. County Council.

Steele, R.C. (1972). Wildlife conservation in woodlands, *Forestry Commission Booklet 29,* London: HMSO.

Treasury (1972). *Forestry in Great Britain: an interdepartmental cost/benefit study*, London: HMSO.

Wood, R.F. and Anderson, I.A. (1968). Forestry in the British scene, *Forestry Commission Booklet 24*, London: HMSO.

Zehetmayr, J.W.L. (1971). The future of forestry in Great Britain, *Advancement of Science*, **28**, Dec.

Forestry Commission Publications

Below is a selection of our recent issues

Annual Report for the year ended March 1974	£1.05
Report on Forest Research for the year ended March 1974	£1.10
Booklet No 15. Know your Conifers	30p
Booklet No 18. Forestry in the Landscape	25p
Booklet No 20. Know your Broadleaves	£1.00
Booklet No 23. Timber: Your Growing Investment: A comprehensive fully illustrated account of the Commission's activities	32½p
Booklet No 29. Wildlife Conservation in Woodlands	75p
Booklet No 38. Common Trees	11p
Forest Record No 91. Birds and Woodlands	40p
Leaflet No 55. Hydratongs	10p
Leaflet No 56. Grey Squirrel Control	12p

Guides

Bedgebury (Kent) Pinetum and Forest Plots	90p
Dean Forest and Wye Valley	65p
East Anglian Forests	60p
New Forests of Dartmoor	21p
Explore the New Forest	£1.85
Queen Elizabeth Forest Park (The Trossachs)	60p
Forestry Commission Potential for Permanent Tourist Accommodation (obtainable only from Forestry Commission)	2.00

Postage extra

Publications are available from H.M.S.O., 49 High Holborn, London, WC1V 6HB (post orders to P.O. Box 569, London, SE1 9NH) through booksellers, or direct from the Forestry Commission, 231 Corstorphine Road, Edinburgh or Publications Section, Alice Holt Lodge, Wrecclesham, Farnham, Surrey, GU10 4LH.

Guide map to your Forests (48p) pamphlet "See your Forests" and a complete publications list available only from the Forestry Commission.

A Note on Geographical Papers

Reading Geographical Papers are published by the Department of Geography at Reading University and present the research findings and opinions on current planning and geographical problems of persons either working or giving lectures in the Department. The series in its current form was started in January 1973 and each paper is printed in offset litho by George Overs Ltd., in a booklet of A5 size. The series is gaining in popularity; we now have over 90 standing orders, we print some 500 of each paper, and casual sales are increasing quite quickly. One or two large bookshops (eg Dillons, and the Economist Bookshop) are now displaying the papers. Of course, the series is non-profit-making.

We have been fortunate in having a number of well-known authorities in Geography and Planning producing papers in the series, for example, Peter Hall, David Harvey, Brian Goodall, and Lloyd Rodwin, and at present we have some 10 new papers scheduled for next year; 4 new papers have just been produced. In the past, we have attracted some advertising of planning and geography text-books, and we would like to attract more advertising. Our rates, which are subject to negotiation, are £10 for a half-page advert, and £18 for a full page.

Papers in Preparation

Domestic Service in Late Victorian and Edwardian England, 1871-1914 by Mark Ebery and Brian Preston.

Abstract

Despite the economic and social importance of domestic service in Nineteenth and early Twentieth-Century English society, little work has been published to date either of a qualitative or quantitative nature. This paper consists of four major sections. The first section reviews the broad features and growth of domestic service in England over the period 1871-1914, particularly in terms of the changes that occurred in the social and spatial structure of the service. The second section examines in detail the life-style of the servant (household structure, age, sex, marital status, birthplace, recruitment, accommodation, wages, hours of work, duties, problems and prospects). To highlight in quantitative terms the detailed discussion in section two, a sample of urban and rural households has been selected from the unpublished 1871 census enumeration books for Bath, Bolton, Coventry, Hastings, Lincoln and Reading, rural Berkshire and South-East Lancashire. Finally, contemporary attitudes towards service are examined from the standpoint of the social investigator, legislator, master and servant.

GP-38

The Education Perspective of Regional Science and Urban Studies by L. Rodwin.

Abstract

Education in urban and regional studies is proceeding today under four main banners. The first stands for Town, City and Land Use Planning; the second Regional Studies; the third Policy Studies, and the fourth, Urban Studies. Each programme makes somewhat different assumptions on the role of the field. These affect the subject matter and the strategy for development. This paper seeks to show how the programmes differ, and seeks to highlight the exceptional advantages of the Urban Studies approach. The paper's remarks apply to the United States, and are based primarily on personal experience at the MIT.

GP-40

Gravity Models in a Dynamic Framework by A. Rodriguez-Bachiller.

Abstract

This paper concerns the possibility of improving the theoretical basis for gravity-type location models by considering them within the context of the dynamics of location and relocation. Instead of applying such models to equilibrium situations, the emphasis here is on the processes which lead to such equilibrium states. This provides the basis for a speculative hypothesis that although the gravity model appears to represent locational behaviour in general quite well, it is the linear programming model which best fits the behaviour characteristic of relocating activities.

The first chapter reviews the different types of dynamic models that have been used in economics and in urban and spatial analysis in general, so as to provide a framework in which to embed the research. The second chapter discusses the implications of using gravity models in dynamic frameworks by studying the relationships between the gravity and linear programming models. When both of these models are used as location models, these models can be interpreted as models of the same urban activity with different locational behaviour rather than as alternative approaches. In fact, it is suggested that behaviour characteristics of the gravity model 'tends', ceteris paribus, to the optimal behaviour of the linear programming model through time.

Data from Central Berkshire for the 1951-66 period is used to support this hypothesis. In the third and last chapter the consequences of the hypothesis previously discussed are put together into a dynamic Lowry-type model incorporating various ideas from the Systems-Dynamics approach. The model is calibrated with the same data.